# SPATIAL PATTERN DYNAMICS
# IN AQUATIC ECOSYSTEM MODELLING

# Spatial Pattern Dynamics
# in Aquatic Ecosystem Modelling

DISSERTATION

Submitted in fulfilment of the requirements of
the Board for Doctorates of Delft University of Technology
and of the Academic Board of the UNESCO-IHE Institute for Water Education
for the Degree of DOCTOR
to be defended in public
on Monday, June 29, 2009 at 14:00 hours
in Delft, The Netherlands

*by*

**Hong LI**
*born in Nei Mongol, China*

Master of Science in Hydroinformatics with Distinction
UNESCO-IHE Delft, The Netherlands
Master of Science in Hydrology and Water Resources
Hohai University, China

This dissertation has been approved by the supervisor
Prof. dr. ir. A. E. Mynett

Members of the Awarding Committee:

| | |
|---|---|
| Chairman | Rector Magnificus, TU Delft, the Netherlands |
| Prof. dr. R.A.M. Meganck | Vice-Chairman, UNESCO-IHE, the Netherlands |
| Prof. dr. ir. A.E. Mynett | UNESCO-IHE, the Netherlands, supervisor |
| Prof. dr. ir. G. Ooms | TU Delft, the Netherlands |
| Prof. dr. ir. G.S. Stelling | TU Delft, the Netherlands |
| Prof. dr. J. O'Keeffe | UNESCO-IHE Delft, the Netherlands |
| Prof. dr. P. Goodwin | University of Idaho, USA |
| Prof. dr. S. Marsili-Libelli | University of Florence, Italy |
| Prof. dr. ir. H.J. de Vriend | TU Delft, the Netherlands, reserve member |

CRC Press/Balkema is an imprint of the Taylor & Francis Group, an informa business

© 2009, Hong Li

Published by:
CRC Press/Balkema
PO Box 447, 2300 AK Leiden, The Netherlands
e-mail: Pub.NL@taylorandfrancis.com
www.crcpress.com - www.taylorandfrancis.co.uk - www.ba.balkema.nl

ISBN 978-0-415-55897-6 (Taylor & Francis Group)

Dedicated to my parents Huanzhong Li and Shuyun Gao
for always having faith in me

# Abstract

The rapid developments in economic activities, urbanization, and population growth during the past decades have influenced aquatic ecosystems much more than at earlier times in history. Effects of climate change are likely to add to that in future. Consequently, aquatic ecosystems are becoming even more fragile, leading to loss of biodiversity and changes in spatio-temporal distributions of species composition.

Spatial pattern dynamics is one important aspect used to reflect the consequences of internal and external pressures on aquatic ecosystems, and supplies basic information to help understand the behaviour of these systems as a whole. Therefore, research on spatial pattern dynamics is important for the sustainable management of aquatic ecosystems.

The main goal of this research is to reveal the dominant processes and factors contributing to spatial pattern development at various scales, and to explore appropriate modelling approaches that can be used to represent spatial pattern dynamics. Clearly, such modelling is not an easy task. Ecosystems are generally complex because they involve a large number of components, nonlinear interactions, scale multiplicity, and spatial / temporal heterogeneity. A brief description of some of the main processes and factors involved in aquatic population dynamics is provided in this thesis.

It is shown that physically-based numerical simulation models can capture quite well those processes (often related to physical principles like conservation of mass, momentum and energy) that can be described in terms of mathematical equations. Also, it is demonstrated that data-driven modelling techniques can be used for main factor selection and non-spatial population prediction, provided sufficient data is available. In addition, if some processes are at least partially understood while only limited data is available, fuzzy set theory can be invoked for capturing the dominant processes by combining sparse data with domain knowledge, as shown in a case study on algal bloom prediction for Western Xiamen Bay, China.

Over the past decades, many environmental and ecological models have been developed like physically-based numerical simulation, data-driven modelling, discrete cellular automata, individual- or agent-based modelling, etc. In this thesis, the main modelling tools for analysis and simulation of spatial pattern dynamics are reviewed, and new developments (notably the application of multi-agent systems) are added.

A difficulty in developing aquatic ecosystem models that has been well known for a long time already, is the limitation of available ecological data, especially spatially distributed data. In this research, advent of more recent spatial information sources is explored, notably remote sensing images, GIS density maps, and high resolution photographs. Such data can supply information for model calibration as well as reference information for model validation. Moreover, spatial information proves most helpful in selecting the dominant processes to be included in the models, as shown in a number of case study applications.

Ecosystem dynamics can be caused by spatial heterogeneity, stochasticity, discrete events, local effects and physical, biological/ecological properties. A number of modelling approaches with different emphases and different modelling requirements were explored and applied to real world situations in order to increase the understanding of the underlying mechanisms and processes involved. This thesis incorporates different case studies ranging from large scale North Sea algal population simulation to meta-population dynamics in inland lakes and individual-oriented macrophytes' growth dynamics in a small pond.

It is shown that by including spatially distributed inputs, a conventional physically-based numerical model could be enhanced to better represent harmful algal bloom spatial dynamics. Not only the importance of spatially heterogeneous information needs to be emphasized, the simulation of discrete events and biological diffusive processes are also of vital importance. When the behaviour of individual plants and/or local effects are vital to the development of spatial patterns, discrete Cellular Automata (CA) and Individual-Based Modelling (IBM) techniques seem more suitable for representing spatial pattern dynamics than equation-based formulations.

This research demonstrated that Cellular Automata are good in representing discrete phenomena like individual-oriented plant growth which are known to mainly depend on local effects. However, in cases where interacting species coexist in the same local environment and competition among them is taking place within each computational cell, traditional CA becomes limited due to its lack of flexibility, especially when such competition is dominated by individual species properties.

In order to better include the (super)individual properties of particular species and account for the interactions among different species, the use of Multi-Agent Systems (MAS) was explored by integrating different data sources (e.g. GIS data), multi-process formulations, and inter/intra species interactions. Results for macrophyte growth seem to capture well the spatial patterns exhibited on GIS density maps obtained from field measurements.

The results from the MAS model also showed that external factors (e.g. water depth and flow velocity) are to be combined with individual properties of aquatic plants' species in order to correctly represent the spatial pattern dynamics.

The general features of aquatic ecosystems are related to multi-processes, multi-scales, nonlinearities and randomness amongst other factors. There is the need to have a modelling framework that can include both turbulent water motions and aquatic population dynamics. Therefore, a synthesis approach of multi-process, nonlinear feedback systems into a unified modelling framework was developed in this research. This synthesis approach included both continuous processes, conventionally modelled by differential equations, and discrete processes modelled here by a Multi-Agent System approach. Although the feedback from aquatic plant dynamics to the flow pattern was not included yet in this synthesis approach, the results in this research show that integration of different types of models holds the potential of representing the combined features of nonlinearity, randomness and complexity of aquatic ecosystems. Moreover, the synthesis approach developed here enhanced the conventional differential equation based modelling that was already available.

The case studies presented in this thesis are mainly research oriented. In future, more project-based applications are to be explored. Also, impacts from global warming and effects of climate change on aquatic ecosystems should be addressed to facilitate the decision-making process for better management of the aquatic environment.

Due to the complexities of aquatic ecosystems and the difficulties of mathematically formulating biological and ecological knowledge, considerable attention should be paid in future research on Uncertainty Analysis (UA). Collaborating with specific domain experts, performing more detailed sensitivity analyses, acquiring more spatio-temporal data, and obtaining better mathematical formulations for the different processes – all are needed to further advance the modelling capabilities. The environmental hydroinformatics techniques and synthesis framework introduced in this thesis can be seen as first steps towards a next generation systems in aquatic ecosystem modelling.

# Samenvatting

De snelle ontwikkeling van economische activiteit, verstedelijking en bevolkingsgroei gedurende de afgelopen decennia hebben aquatische ecosystemen aanzienlijk meer beïnvloed dan de ontwikkelingen gedurende lange tijd daarvoor. Naar alle waarschijnlijkheid zullen effecten van klimaatverandering dit in de toekomst nog verder versterken. Het gevolg hiervan is dat aquatische ecosystemen nog meer te verduren krijgen met als te verwachten resultaat een verder verlies aan biodiversiteit.

Het dynamisch gedrag van ruimtelijke patronen wordt vaak als maat gehanteerd voor de gezondheid van natuurlijke watersystemen en verschaft fundamenteel inzicht bij onderzoek naar veranderingen die zich binnen deze systemen voordoen. Aquatische ecosystemen omvatten een groot aantal processen dat plaatsvindt over een scala aan tijd- en lengteschalen. Niet-lineaire interacties, natuurlijke variabiliteit, incidentele gebeurtenissen, en terugkoppelmechanismen tussen biologische, chemische en fysische processen leiden vaak tot complexe verschijnselen.

Het hoofddoel van dit onderzoek is om na te gaan welke processen en factoren het meest van belang zijn voor de ontwikkeling van ruimtelijke patronen en om te onderzoeken welk type modellen zich het best lenen om de dynamica daarvan te beschrijven. In veel gevallen vraagt dit om een combinatie van conceptuele en numerieke modellen met meetgegevens en specifieke kennis vanuit de onderliggende vakgebieden, gebruik makend van de mogelijkheden die computersystemen en informatie- en communicatietechnologie bieden. Dit is het werkgebied van de discipline "Environmental Hydroinformatics".

Het ontwikkelen van dergelijke gecombineerde modellen is geen sinecure. Niet alle processen laten zich immers op eenzelfde wijze beschrijven in wiskundige formuleringen. Zo bestaan er enerzijds numerieke modelsystemen zoals Delft3D met zijn DELWAQ Open Proces Library, en anderzijds specifieke technieken uit het vakgebied van de kunstmatige intelligentie, zoals Artificiële Neurale Netwerken. In dit proefschrift wordt aangetoond dat op data gebaseerde modellen (ANNs) zeer goed in staat zijn om de belangrijkste factoren van een bepaald proces te selecteren, mits uiteraard voldoende data beschikbaar zijn. Een ander voorbeeld betreft het combineren van meetgegevens en vakkennis middels Fuzzy Set theorie waarmee dominante processen kunnen worden geïdentificeerd, zoals blijkt uit een onderzoek naar algenbloei in Western Xiamen Bay, China.

Gedurende de afgelopen decennia zijn vele modelsystemen ontwikkeld om het gedrag van aquatische ecosystemen te beschrijven. Naast bovengenoemde numerieke simulatie- en data-georienteerde modellen kunnen bovendien worden genoemd discrete cellulaire automaten, modellen op basis van individueel gedrag, etc. alsmede. In dit proefschrift wordt een overzicht gegeven van de huidige beschikbare modelsystemen voor de analyse en simulatie van ruimtelijke patronen, terwijl eigen nieuwe ontwikkelingen (met name de toepassing van "multi-agent systems") worden toegevoegd en de integratie en combinatie van deze verschillende benaderingen in de vorm van een "sythese-model" wordt uiteengezet.

Een limiterende factor die al geruime tijd bestaat bij het modelleren van aquatische ecosystemen is de beperkte beschikbaarheid van ecologische meetgegevens, in het bijzonder met betrekking tot de ruimtelijke verdeling van de verschillende grootheden. In dit proefschrift wordt uitgebreid ingegaan op de mogelijkheden die de huidige informatietechnologie kan bieden, zoals het gebruik van Remote Sensing beelden, Geografische Informatie Systemen, en hoge resolutie fotografie. De gegevens die hieruit worden verkregen kunnen worden gebruikt voor het selecteren van de belangrijkste processen die in de modelsystemen dienen te worden meegenomen, zoals aangetoond in een aantal praktijktoepassingen.

Er zijn veel factoren die de dynamica van ecosystemen beïnvloeden, waaronder ruimtelijke variaties, stochastische effecten, discrete gebeurtenissen, locale condities en specifieke biologische/ecologische eigenschappen. Verschillende model aanpakken met verschillende invalshoeken en vanuit verschillende toepassingsgebieden zijn in dit onderzoek vergeleken op basis van hun praktische bruikbaarheid. In dit proefschrift worden verschillende praktische toepassingen bestudeerd, variërend van grootschalige algenbloei modellering op de Zuidelijke Noordzee, tot het dynamisch gedrag van meta-populaties in meren, alsmede de individuele groei van waterplanten in een vijver.

Door de ruimtelijke verdeling van de beginvoorwaarden mee te nemen, kon een bestaand numeriek model voor het voorspellen van de dynamische eigenschappen van plaagalgen worden verbeterd. Naast de ruimtelijke verdeling is het ook van belang om specifieke gebeurtenissen in de tijd ('events') op een juiste manier mee te nemen. In dit onderzoek is nagegaan welke alternatieven er zijn om variaties in ruimte en tijd te combineren in eenzelfde modelsysteem.

In geval het gedrag van individuele planten en locale effecten van doorslaggevend belang zijn voor de ontwikkeling van ruimtelijke patronen, dan lijken Cellulaire Automaten (CA) en 'Individual-Based Modelling'

(IBM) dan wel 'Agent-Based Modelling' (ABM) beter geschikt om ruimtelijke patronen te ontwikkelen dan formuleringen gebaseerd op algemeen geldende veldvergelijkingen. In dit onderzoek wordt aangetoond dat CA heel goed kan worden gebruikt voor het modelleren van groei en afsterven van individuele waterplanten die met name van locale condities en interacties afhankelijk zijn. Echter, in geval meerdere soorten gelijktijdig in eenzelfde gebied wedijveren om voedsel, dan ligt de toepassing van CA minder voor de hand vanwege de beperkte mogelijkheden om specifieke soorteigenschappen binnen eenzelfde rekencel te representeren.

Om voor dergelijke toepassingen betere modelsystemen te ontwikkelen die de interacties tussen verschillende soorten wel kunnen meenemen, zijn in dit onderzoek de mogelijkheden nagegaan van Multi-Agent Systems (MAS). Daarbij is gebruik gemaakt van de combinatie van verschillende informatiebronnen (waaronder GIS-bestanden), procesbeschrijvingen en de interactie tussen soorten onderling en met hun omgeving. De resultaten van een praktische toepassing van de groei van twee soorten waterplanten in eenzelfde gebied komen goed overeen met waarnemingen opgeslagen in een GIS systeem. Daarbij bleek tevens dat naast de specifieke eigenschappen van soorten, ook externe factoren als waterdiepte en lichtdoordringing van groot belang zijn bij de evolutie van ruimtelijke patronen. Deze kunnen echter relatief eenvoudig uit reeds bestaande modelsystemen worden verkregen.

Aquatische ecosystemen laten zich kenmerken door de interacties tussen meerdere processen op meerdere tijd- en lengteschalen, met niet-lineaire uitwisselingen en toevallig optredende gebeurtenissen. Dit vraagt om een modelleer raamwerk dat zowel de turbulente waterbeweging kan representeren als het dynamisch gedrag van de verschillende plant- en diersoorten. Zo een raamwerk is in dit onderzoek ontwikkeld, waarbij naar een synthese is gestreefd tussen het modelleren van continue processen op basis van partiële differentiaalvergelijkingen en discrete processen middels een Multi-Agent System benadering. Hoewel de terugkoppeling tussen de groei van waterplanten op het stromingspatroon nog niet in de huidige synthese is meegenomen (de basis formuleringen zijn wel bekend), laten de resultaten zien dat het hier ontwikkelde synthese model mogelijkheden biedt om de relevante factoren te combineren. Daarbij is de hier ontwikkelde aanpak zodanig opgezet, dat deze in staat is om biologische en ecologische uitwisselingsprocessen direct in te passen in bestaande modelsystemen, zoals Delft3D dat hier is gebruikt.

De toepassingen die in dit proefschrift staan beschreven waren met name gericht op onderzoek naar verbeterde modelvorming. Het is uiteraard wenselijk om de praktische toepassingen verder uit te breiden. Dit geldt ook voor het ontwikkelen van modelsystemen waarmee effecten van klimaatverandering kunnen worden bestudeerd om te komen tot een beter beheer van het aquatisch milieu.

De complexiteit van aquatische ecosystemen en de moeilijkheid om biologische en ecologische kennis in wiskundige formuleringen te gieten vragen expliciet aandacht en verder onderzoek naar het omgaan met onzekerheid. Samenwerking met deskundigen uit verschillende vakgebieden is daarbij onontbeerlijk, evenals het verkrijgen van ruimtelijke meetgegevens over langere termijn, en het afleiden van betere wiskundige formuleringen voor de verschillende processen – deze zijn alle nodig om tot betere modelvorming te komen. De environmental hydroinformatics techniques en het synthese model die in dit proefschrift zijn geïntroduceerd kunnen gezien worden als bijdragen op weg naar de ontwikkeling van een volgende generatie softwaresystemen voor aquatische ecosysteem modellering.

# Acknowledgement

This PhD thesis is a result of the research programme in Environmental Hydroinformatics carried out in close collaboration between UNESCO-IHE Institute for Water Education, Delft University of Technology (TUDelft) and WL|Delft Hydraulics, now part of Deltares. The research was funded by the Department of Strategic Research & Development (S&O) of Delft Hydraulics, as part of its strategic research & development programme. I would like to express my sincere appreciation to all organizations and persons that directly or indirectly stimulated me to conduct this research.

Many thanks are due to the Nuffic scholarship 2003-2005 which enabled me to travel to the Netherlands to start my MSc study in Hydroinformatics at UNESCO-IHE. During my entire time in the Netherlands, I have been able to widen my knowledge and vision by learning from professors, colleagues and friends from all over the world.

To Prof. Arthur Mynett my appreciation and gratitude go much further than what I can write here. It was you giving me the chance to start this PhD research, and it was you supervising me systematically and tirelessly through both my MSc and PhD research during the past five years. You have been sharing your vision and wisdom, trusting my capability in doing research and guiding students, giving me freedom to develop new ideas and allowing me to attend quite a number of international conferences as part of this research. Thank you very much for your continued support; without you, this study would not have been finished successfully.

During the PhD research, I have had the opportunity to work together with many experts in a wide number of application areas, including hydrodynamics, water quality, ecology, artificial intelligence, geographical information systems and remote sensing. Thank you all for these great experiences. Thanks to Ir. Anouk Blauw, Dr. Hans Los and all IVM experts, as well as to Mr. Mijail Arias and Ir. Gerald Corzo for the excellent teamwork we had. Thanks to Ir. Ellis Penning and Mrs. Hui Qi for the case study we did together on aquatic plant dynamics modelling. Many thanks, Ellis, for your kindness and your biological knowledge as well as for allowing me to participate in field measurements in Lake Veluwe: a wonderful learning experience! Thank you, Ir. Leo Postma and Mr. Qing Hua Ye, for helping me in the use of the DELWAQ Open Process Library.

Other benefits from the excellent team work we enjoyed during this research were the possibility to share data and information. I acknowledge very much the data suppliers of this research: the Remote Sensing records from IVM (Institute for Environmental Studies) at the Vrije Universiteit Amsterdam, the GIS maps for Lake Veluwe from Rijkswaterstaat IJsselmeergebied, the meteorological data from the Royal Netherlands Meteorological Institute and the high resolution photographs from the Freshwater Ecology Group in Deltares (WL|Delft Hydraulics). Without the spatially distributed data from all above organizations and research groups, this research could not have been carried out successfully.

I also acknowledge the kind collaboration with Dr. Ann van Griensven from UNESCO-IHE. You helped me in correcting my papers, guiding me in some of my ideas and trusting my capability of doing research. And thank you for being my friend. My pure-hearted gratitude also goes to all other staff members of Hydroinformatics at UNESCO-IHE: Prof. Roland Price, Prof. Dimitri Solomatine, Dr. Ioana Popescu, Dr. Andreja Jonoski, Dr. Yunqing Xuan, Dr. Biswa Bhattacharya and Dr. Zoran Vojinovic, as well as many other professors, lecturers, and staff members at UNESCO-IHE. I really value your contributions to my education.

I have been very fortunate to have many friends who always supported me and shared nice times together with me. Thank you my friends: Mijail Arias, Shengyang Li, Yuqing Lin, Yenory Morales, Ann Sisomphon, Hui Qi, Zhuo Xu, Taoping Wan, Qing Hua Ye, Migena Zagonjolli, Xuan Zhu. Many thanks also go to all staff members and other colleagues in the S&O department of Deltares. Particularly, Mrs. Jitka van Pomeren-Velkova and Mrs. Frances Kelly, I give my special thanks to you, for always helping me in so many things in my work and for creating a very comfortable and convenient working environment, giving me advise on many things in life, and for being my good and wise friends. Thanks also to all other friends who I haven't mentioned but who have helped me and encouraged me.

I would also like to express my gratitude to all committee members for reading my thesis and giving me great comments for improving my thesis.

My greatest thanks go to my family. Thank you grandma for loving me, cherishing me and for your strong personality. Thank you papa and mama for your endless love, for your encouragement and for always believing in my fantasies. Thank you Ziyi for being such an adorable and understanding son, for facing the difficulties together with me in the Netherlands, for the happiness you bring to me. Thank you, Wen, as my husband for always supporting me. Thank you, Li Li my sister and Li Gang my brother for your love and for being around our parents when I cannot be there. Special thanks to my parents in law for your love and for treating me as your own daughter and for helping me in taking care of my son. Thanks to all my uncles, aunties and other relatives for being proud of me and for loving me.

# Contents

# Chapter 1

# Introduction

## 1.1 Research scope

### 1.1.1 Background

Aquatic ecosystems are often characterized by dynamic multi-scale processes in which aquatic organisms live and interact with their environment (Jørgensen and Bendoricchio, 2001). They are in general complex due to the wide range of time and space scales involved and nonlinear due to the coupling and variability of biological, chemical and physical processes in a turbulent aquatic environment. The rapid developments in economic activities, urbanization, and population growth during the past decades have influenced aquatic ecosystems much more than at earlier times in history (Alberti et al., 2003). Effects of climate change are likely to add to that in future as well. Consequently, aquatic ecosystems are becoming even more fragile, leading to loss of biodiversity and changes in spatio-temporal distributions of species composition.

Hence, at present more and more attention is being paid to protecting the aquatic ecosystem. The Millennium development goals (United Nations, 2008) have put emphases on ensuring environmental sustainability through improved environmental management. The EU Water Framework Directive (United Nations, 2008) requires that all inland and coastal waters within defined river basin districts must reach the "good" status by 2015, and defines how this should be achieved through the establishment of environmental objectives and ecological targets for surface waters (Griffiths, 2002). In this context, expert knowledge from researchers and scientists in different fields are needed for decision makers to improve their aquatic environmental management and stimulate future sustainable development.

As the most vital primary producers in aquatic ecosystems, aquatic plants, including algae and macrophytes, live in such a complex dynamic environment where they supply food to other living organisms. They are themselves a physical environment in which fish and other aquatic species find their own habitats. Their dynamic distributions in space and time are crucial aspects in aquatic ecosystems. Dynamic changes occur all over the system at all scales from chemical reactions within seconds, to growth in day-night cycles, to population density changes over seasons, etc. Such dynamics not only develop with their own rhythm, but also are disturbed and interrupted by other factors.

Aquatic plants' dynamics are studied by many biologists, ecologists, and scientists from many fields by means of laboratory research, field measurements, as well as conceptual and mathematical modelling and simulation. The research scales vary from individuals to populations, to communities, and to entire ecosystems. Population behaviour is the result of groups of individuals interacting with each other and with their environments, whereas phenomena at the scales of communities and ecosystems can evolve from population levels.

During the last a few decades, the research focus in aquatic population dynamics covered many aspects including species growth and interaction among different species (Minns et al., 2000; Berger et al., 2008), reaction to environmental changes (Gardner and Engelhardt, 2008; Jeltsch et al., 2008), as well as recently researched feedback mechanisms to water quality, fluid flow properties and to other living organisms (Uittenbogaard, 2003; Baptist, 2005). The importance of spatial patterns in ecology has long been recognized (Watt, 1947). Many researchers in landscape ecology have worked on the modelling of spatial dynamics of terrestrial vegetation, but hardly on the dynamics of aquatic plants' spatial distribution and spatial extension, although spatial pattern dynamics of aquatic plants are very important indicators for quantifying habitat complexity (Chen et al., 2002). Spatial pattern dynamics reflect the consequences of internal and external changes in the system (Dale, 2004), and also supply basic information to research on the dynamics of other organisms in the water bodies. Hence spatial pattern dynamics (e.g. patch size, patch locations, patch spacing, patch density, and patch species composition) can help understand the behaviour of other components of the aquatic ecosystem. Therefore, the research on spatial pattern dynamics is very important for sustainable management of the aquatic environment.

Models are simplified representations of the real world (Mynett, 2002; Price, 2006). They can be very helpful in simulating aquatic plant growth behaviour and spatial expansion, and their results help to obtain a better understanding of the underlying mechanisms. Modelling results also can serve as information for decision makers to mitigate harmful events due to population outbreaks, and to reduce the economical and ecological costs. However, such modelling is not an easy task. Ecosystems are generally considered among the most complex because they are characterized by a large number of diverse components, nonlinear interactions, scale multiplicity, and spatial heterogeneity (Wolfram, 2002). The spatial scales at which spatial patterns are formed are affected by the underlying processes, which vary from individual to landscape scales (Levin, 1992). For each feature at a particular scale, a different descriptive model could be written.

Scales also refer to the scale of observation, the temporal and spatial dimensions at which particular phenomena are observed (Peterson and Parker, 1998), etc. The scale of observation is a fundamental factor in our descriptions and explanations of the natural world. Observation scales define a minimum level of detail at which processes should be included in the model (Wiegand et al., 2003). Observed patterns can also determine the degree of aggregation of biological information. As data availability often is an overarching problem in ecological modelling, Remote Sensing (RS) images and GIS techniques are increasingly being used, which still need to be further explored when considering population spreading and outbreaks. The use of remote sensing images in pre-processing and post processing of numerical model simulations can be found in many papers; however, it is not widely explored how to use RS images in the modelling process itself (e.g. through data-assimilation), which can be extremely useful when long time continuous series of images are available. High resolution photos can supply detailed two dimensional pattern changes through time, and can be used as data sources for modelling by combining them with GIS techniques, as shown in this thesis.

### 1.1.2 Available approaches

In the past decades, many environmental and ecological models, including physically-based numerical modelling (Jørgensen and Bendoricchio, 2001; Postma, 2007; Los et al., 2008), artificial intelligence (Chen et al., 2006; Li et al., 2006b; Li et al., 2007b), discrete cellular automata modelling (Silvertown et al., 1992; Balzter et al., 1998; Mynett and Chen, 2004; Li et al., 2006a; Li et al., 2008b) agent or individual based models (Morales et al., 2006; Hovel and Regan, 2008), as well as the integration of different modelling approaches (Giusti and Marsili-Libelli, 2006; Milbradt and Schonert, 2008), have been used increasingly in the modelling of aquatic population dynamics. The modelling approaches have been used not only for prediction, but also more importantly for obtaining a better understanding and explanation of the underlying mechanisms. In general, the modelling approaches can be categorized into three types: equation-based, discrete modelling approaches and agent based models. These are introduced briefly hereafter.

Physically-based modelling is often based on conservation principles for a homogeneous medium that can be conveniently expressed in terms of (partial) differential equations, which has become one of the main modelling approaches for fluid flow and water quality transport phenomena, but also for classical population dynamics modelling. Normally, models of such kind are deterministic, representing equilibrium conditions for ecosystems with averaged parameters. This type of modelling system is composed of coupled state variables, each describing a continuous quantity and evolving smoothly in a time continuum, such as population density or averaged biomass. However, as mentioned above, any real system often exhibits considerable variability in space and time and hardly ever reaches equilibrium conditions (Jørgensen and Bendoricchio, 2001). Recent work has shown the importance of spatial and temporal dynamics on aquatic population behaviour. Emergent theories have shown that global patterns over space and time may evolve from local interactions of adaptive individuals with each other and with their environment. Purely physically-based models often become insufficient for the representation of spatial pattern dynamics including spatial patchiness and species interactions (Shnerb et al., 2000) due to the underlying assumption of spatial and temporal homogeneity, ignoring individual differences and local interactions. There is a clear need to improve further the existing numerical modelling approaches by adding more spatial heterogeneity, or by incorporating alternative approaches that can represent different processes beyond spatially continuous density distributions. Therefore, this research aims to explore such alternative approaches and possible solutions to improve current modelling practice and achieve a better understanding and better simulation of spatial pattern dynamics of aquatic populations.

Apart from differential equations that can represent complex behaviour in homogeneous media from relatively simple formulations (Simon, 1996), there are two more paradigms that can be used for population dynamics modelling: the discrete paradigm and the agent-based paradigm (Ratz et al., 2007). Among the discrete paradigms, Cellular Automation (CA) models seem to be the most common to deal with spatial variation and local interactions. CA has become increasingly popular in ecosystem modelling since it uses simple rules to form global complexity. It deals with small patches on a discrete grid, considering local interactions between populations and their environment, hence having great potential of representing e.g. population growth spreading mechanisms and spatio-temporal variation. Although CA-based computational models cannot be used to precisely forecast events over the long-term (Parrott and Kok, 2006), they can contribute to ecology by providing useful illustrations of the nature of ecosystem dynamics and of the mechanisms that give rise to unexpected events (e.g. catastrophe, population outbreaks).

The differences between CA type models and continuous models mainly show up in two aspects: discrete versus continuous approach, and heterogeneity versus homogeneity, as shown in (Shnerb et al, 2000) and (Minns et al., 2000). CA can achieve global behaviour derived from local interactions, and therefore is good in describing phenomena that depend primarily on the states of each cell and its immediate neighbours. The interactions between species are obtained from neighbouring cells where each cell represents one species with a certain state. The performance of such types of models depends on local rules and sometimes initial conditions. Local rules are generally deterministic in nature and often – but not necessarily – apply to each cell. Initial conditions may vary, depending on the knowledge of the particular process and data availability.

The limitation of available knowledge and data can lead to difficulties in setting up the local rules. In this research, alternative data sources and various types of cellular automata models are studied in order to overcome these difficulties and achieve suitable and robust formulations for modelling aquatic population dynamics.

In most of the real cases, interacting species coexist in the same local environment and competitions amongst them are mainly due to their own growth properties in the local environment. Under such conditions, traditional CA becomes limited due to the lack of flexibility in its own concept and structure. Besides, cellular automata do not necessarily describe individuals at the species level, but more at aggregated conditions (Grimm and Railsback, 2005). To overcome these shortcomings, the relatively new concept of Multi-Agent Systems (MAS) have a better way of describing natural phenomena (Ferber, 1999), since this type of modelling approach allows aquatic spatial pattern dynamics to emerge from the behaviour of individuals or groups of individuals and their interactions, which gives more insights into their growth, interaction and spreading mechanisms.

MAS models can be also based on discretizing the modelling domain into either structured grids or unstructured cells that become the computational background. Each agent has its own position in the background environment with its own properties and behaviour. There can be interactions amongst different agents, which are under certain constraints from their surrounding environment. Many different types of agents can coexist in each background cell and their competition can be both within the cell as well as within the neighbouring regions of the cell, depending on the individual properties of each agent. The MAS approach is flexible and robust. However, sometimes computations can be expensive, which becomes less and less a problem because of the rapid development of computational power. This research explores the use of Multi-Agent Systems to the study of aquatic plants' spatial pattern dynamics. New measurement techniques, like time series of photos, remote sensing images, GIS maps etc., together with in situ measurements, can be conveniently used in approaches like MAS to obtain a better representation of the spatial heterogeneity in aquatic population growth.

## 1.2   Objectives and research questions

There are still many gaps in understanding the mechanisms and capturing the proper spatial pattern dynamics in aquatic ecosystem modelling. Application of alternative modelling paradigms may provide more insight into understanding the mechanisms behind unknown phenomena like the formation of patchiness and the simulation of population outbreaks. Therefore, this research tries to combine available scientific insight from aquatic ecosystem theory and measurement data with hydroinformatics tools and technologies, with the aim to explore alternative modelling approaches that can achieve a better understanding and simulation of spatial and temporal patterns of aquatic ecosystem dynamics.

The main research question is: how to represent spatial pattern dynamics in aquatic ecosystem modelling by the combination of models, knowledge and data? This leads to several component questions:

• What is the importance of spatial pattern dynamics in aquatic ecosystems?
• What are the factors and processes involved in spatial patterns of aquatic populations and how to find the dominant factors and processes?
• Which kind of data and information are required to enable different modelling approaches and what is the data availability?
• How does a conventional modelling approach simulate spatial pattern dynamics of aquatic populations and how to enhance conventional models?
• Can we use discrete paradigms and multi-agent systems for aquatic plant dynamics modelling and how to develop such type of models?
• Is it possible to develop a synthesis framework of multi-process ecohydraulics system modelling and how to do this?
• What are future uses of the methods developed in this research in practice?

Hence, the main objective of this PhD research is to explore alternative modelling approaches in representing spatial pattern dynamics of aquatic populations in aquatic ecosystems, which depict many of the features of a complex system over a wide range of scales. There are many detailed objectives for the realization of the main objective:

• To have a good understanding of the mechanisms and processes related to aquatic population spatial pattern dynamics;

- To analyze different types of data including Remote Sensing images, GIS maps, high resolution photographs and in situ observations as well as laboratory data;
- To describe the state-of-the-art of different modelling techniques;
- To learn from conventional modelling approaches and find ways to enhance conventional models;
- To investigate the use of cellular automata based models in the presentation of spatial pattern dynamics of aquatic plant growth;
- To explore the use of multi-agent systems in aquatic ecosystem modelling in order to exhibit spatial pattern dynamics of multi-species;
- To incorporate different types of data sources into alternative modelling approaches in order to achieve a better estimation of aquatic population spatial pattern dynamics;
- To investigate the applicability of a synthesis framework for multi-process ecohydraulics non-linear systems based on above modelling experiments and expert knowledge.

The synthesis of different modelling techniques is somehow difficult since this does not only depend on the experiments done with those approaches, but more importantly, also depend on both the properties of the modelling tools and the spatial and temporal characteristics of the ecosystems that are studied. The problems of spatial and temporal scales in ecological modelling have been identified and studied by many biologists and modellers like (Levin, 1992; Schneider, 1994; Jørgensen and Bendoricchio, 2001; Mynett and Chen, 2004; Parrott and Kok, 2006). By combining modelling approaches with expert knowledge and available data, the applicability of a synthesis of various modelling approaches like physically-based modelling and multi-agent-based modelling, is investigated in this research.

Finally, yet importantly, the research needs to serve as the support for ecosystem sustainable development and management. Therefore, research-oriented real case applications will be implemented in this research in order to supply potential supports in decision making processes for building and maintaining healthy ecosystems in aquatic environment.

## 1.3   Thesis outline

The thesis is composed of nine chapters, which can be grouped into three parts in addition to introduction and conclusions (Figure 1-1):

Part I deals with the state-of-the-art of theoretical knowledge and the available modelling techniques. This part includes Chapter 2 and Chapter 3.

Chapter 2 gives the biological and ecological aspects in aquatic ecosystem dynamics and the factors and processes that lead to the spatial-temporal dynamics of a population. This serves as the general domain knowledge in the alternative modelling tools' development.

Chapter 3 addresses the state-of-the-art in spatio-temporal population dynamics modelling for aquatic ecosystems, which includes the conventional (partial differential) equation-based modelling, discrete modelling and agent-based modelling approaches. This chapter gives detailed information about various modelling approaches and shows the past research on different approaches. This chapter leads to the emerging research needs.

Part II focuses on the development and real case applications of different modelling techniques in representing proper spatial pattern dynamics in aquatic ecosystem modelling. The modelling approaches include data-driven techniques, differential equation based models, Cellular Automata and Multi-Agent Systems. This part is composed of Chapters 4, 5, 6 and 7.

Chapter 4 presents how to use statistical methods and data-driven techniques for the selection of main factors in aquatic population models with large numbers of variables involved in the aquatic ecological processes. It shows the importance of main factor selection and the power of using data-driven techniques also for predicting population outbreaks like algal bloom events. Chapter 5 describes the details of conventional differential equation based models. Remote Sensing images are introduced and discussed in this chapter in order to show that conventional modelling approaches can be enhanced by adding more spatially heterogeneous information. It demonstrates the importance of spatial dynamics in aquatic ecosystem modelling. Chapter 6 summarizes different types of Cellular Automata (CA) models. A traditional fine scale CA is developed to model water lily growth in a small pond based on time series photos and the biological knowledge. Chapter 7 explores the use of Multi-Agent Systems (MAS) in the modelling of macrophytes' pattern development. This chapter shows the power of a MAS model in representing the spatial pattern dynamics in the research area of Lake Veluwe in the Netherlands.

Part III investigates the applicability of a modelling framework for a synthesis of multi-processes for aquatic spatial pattern dynamics modelling (Chapter 8). This framework synthesizes physically-based models with an aquatic plant growth model based on Multi-Agent Systems, incorporating theoretical knowledge with different types of measurements. This modelling framework is implemented in the Delft3D Open Process Library and a case study is carried out to verify its applicability.

Finally, Chapter 9 gives the conclusions and recommendations of this research.

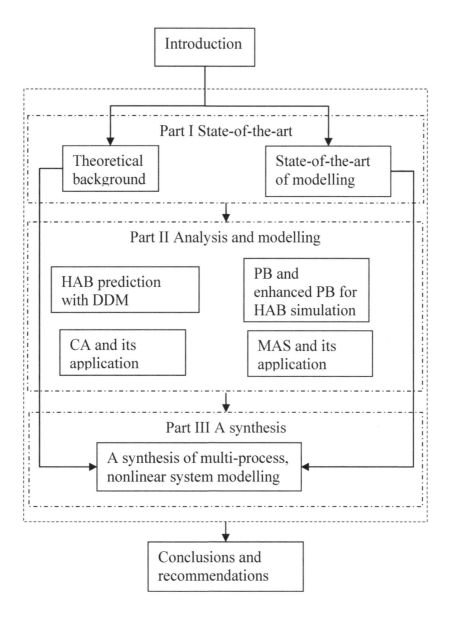

Figure 1-1 Thesis outline
*DDM: Data-Driven Model; PB: Physically-Based Model;*
*CA: Cellular Automata; MAS: Multi-Agent-Systems;*
*HAB: Harmful Algal Bloom*

# Chapter 2

# Theoretical background

## 2.1   Introduction

Aquatic ecosystems in lakes, sea and other water bodies, contain many different species ranging from bacteria, phytoplankton, aquatic plants/macrophytes to fish (Brönmark and Hansson, 2005). Water bodies like sea, rivers, estuaries and lakes provide the physical, chemical and biological conditions for species to adapt. Therefore, aquatic population dynamics is not only influenced by its own properties but also by environmental factors such as water motion, nutrient availability, sediment transport as well as meteorological conditions (e.g. light and temperature) (Brönmark and Hansson, 2005). This chapter focuses on the description of general features of aquatic populations and how aquatic populations interact with their environment. Firstly, section 2.2 briefly describes aquatic ecosystems, followed by the introduction of the concept of population dynamics in section 2.3 that summarizes the processes and factors which interact with populations. Section 2.4 gives a more detailed explanation about what are the specific factors and processes involved and how these factors and processes influence aquatic population dynamics. A short summary is given in section 2.5.

## 2.2   General description

An aquatic ecosystem is a group of interacting organisms depending on one another and their water environment for nutrients, shelter and other things (e.g. $O_2$). Aquatic ecosystems are responsible for a large proportion of the planet's biotic productivity (nearly half of the global net primary production occurs in marine ecosystems conducted by *phytoplankton* (Field et al., 1998)). There are several different types of water bodies forming different types of aquatic ecosystems: oceans, seas, rivers, estuaries, lakes and ponds, and wetlands, etc. They have common features as well as different characteristics. In general, when compared to terrestrial ecosystems, aquatic ecosystems are special in several different ways. Firstly, organisms in aquatic systems must be able to survive partial to total submergence. Water submergence has an effect on the availability of atmospheric oxygen, which is required for respiration, and solar radiation, which is needed in photosynthesis. Secondly, some organisms in aquatic systems critically depend on dissolved substances in their immediate environment. These conditions have caused these forms of life to develop physiological adaptations in order to deal with such situation. Thirdly, aquatic ecosystems are limited by nutrients like phosphorus and iron, etc., especially in fresh waters (less limited by nitrogen) (Brönmark and Hansson, 2005). Lastly, aquatic ecosystems are generally cooler than terrestrial systems; this condition limits metabolic activity, which is generally controlled by temperature. However, there are also big differences between various aquatic ecosystems in terms of physical and biological conditions. Aquatic ecosystems can be categorized into marine and freshwater ecosystems due to the concentration of salt in water, and they can be divided into physically driven systems such as rivers, sea surface, and biological dominated such as most of the ponds and lakes, coral reefs, etc. The aquatic ecosystems considered in this research are: sea (mainly sea surface), coastal waters and estuaries as the main aquatic environment for phytoplankton analysis and modelling; lakes and ponds as the environmental frame for macrophytes. The theoretical background given below focuses mainly on above mentioned aquatic ecosystems.

As a science of the relationships between organisms and their environments, ecology can be catalogued by scales at various levels: (i) at individual level, which explores how the environmental conditions influence particular species leading to specific mutations and eventually to the adaptation of the organism to the prevailing conditions; (ii) at population level, which is on how populations adapt to a changing environment in terms of population growth rates, population viability analysis, population genetics, metapopulation analysis, population outbreaks, etc; (iii) at community level, which includes interspecific interactions and competition, environmental impact survival, etc; (iv) at ecosystem level, which includes the overall energy balance, propagation of matter, nutrient flow, pollution impact; and (v) at biosphere level, e.g. climate change, etc.

Each of these levels is influenced by its environment at corresponding spatio-temporal scales. In an aquatic ecosystem, all species from bacteria to primary producers to fish and other aquatic animals belong to the community are all subjected to the external environment. In this research, the focus is mostly on populations and metapopulations (a population of many local populations - (Levins, 1970; Hanski and Simberloff, 1997)) as important components of aquatic ecosystems.

## 2.3   Population life-system

A population is defined as a collective group of organisms of the same species (Jørgensen and Bendoricchio, 2001). Each population has several characteristic properties, such as population density, natality, mortality, age distribution, dispersion, growth forms and others. A population is changing through time and space based on the rate of growth, death, migration, etc. Besides its own intrinsic properties, a population is also influenced, sometimes even dominated, by available resources (e.g. food, shelter, and breeding space), enemies (e.g. predators, parasites) and ambient conditions (e.g. light, temperature, hydrodynamic conditions, etc.) which together form the components of a population system (or a life-system) defined by Clark et al. (Clark et al., 1967) as a population with its effective environment.

Population systems are extremely complex, involving numerous processes and factors interacting between one and another, which are not fully understood but vital for mitigating the consequences of ecosystem unbalance and population outbreaks.

The factors (the components in the life-system) and the processes (the events changing population systems) interact with each other and form the dynamics of the population system (Sharov, 1992). Factors affect the rate of change of processes (Figure 2-1) and processes change the value of factors (Figure 2-2).

The processes which are related to the population dynamics include physical (e.g. transportation), chemical (e.g. photosynthesis), biological (e.g. growth, mortality, feeding, species competition, etc.), ecological processes (e.g. habitat availability, etc.) and evolutionary processes (not the main concern in this research). The processes listed in Figure 2-1, and Figure 2-2 are the ones directly linked to the population dynamics from the population level point of view. Some of them are considered as spatial processes (e.g. spatial extension, dispersion, transportation, etc.), and some others are considered none spatial (e.g. growth, mortality etc).

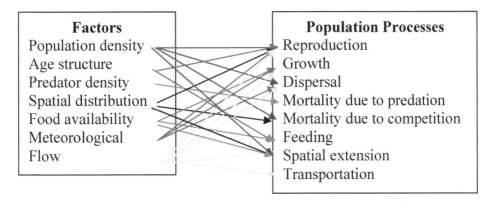

Figure 2-1 Factors affect the processes in population systems

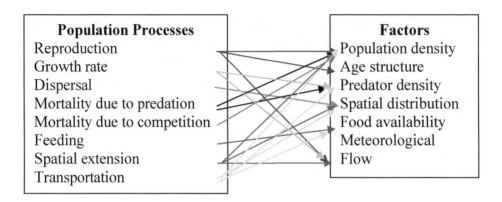

Figure 2-2 Processes change the values of factors in population systems

## 2.4   Processes and factors influencing aquatic population dynamics

In aquatic population systems, which include the population and its effective environment, the populations are interacting among themselves and at the same time with their environment. The complexity related to this system includes different processes and many factors. Aquatic population dynamics are spontaneously determined by both abiotic (e.g. climate, topography, latitude and altitude, flow motion and substances in water bodies) and biotic factors (intraspecific interactions and interspecific interactions respectively) as well as by their own biological growth. At the same time, they give rise to feedback to their external environment they live in (e.g. by reducing light penetration, etc.).

Within the complexity of interacting processes and factors, it is important to note that different species can adapt differently to their environment, causing some to be successful and perhaps become dominant in time and/or space. Both the environment and the characteristics of aquatic populations, including abiotic and biotic factors, determine their existence and spatio-temporal development. Hence, species composition in a water body becomes closely related to the characteristics of the coastal waters, lakes or ponds investigated, including nutrient status, climate conditions, composition of grazers, acidity levels, as well as flow patterns, etc.

### 2.4.1  Impact of abiotic factors

Being the main types of aquatic populations considered in this research, aquatic plants may be divided into macrophytes, phytoplankton and periphytic (substrate-associated) algae (Brönmark and Hansson, 2005). These in general have similar resource requirements, for example, phosphorous, nitrogen, carbon dioxide and light, etc. Besides, the meteorological conditions (wind, rainfall, temperature, etc.), hydrodynamic characteristics (current, stratification, turbulence, etc.) can influence or even determine their growth dynamics. But they may also have different solutions and adaptations to their environment and the changes therein.

### Currents and turbulence

Currents and turbulence obviously have a large impact on pelagic animals and plants with limited swimming capacity. In addition, they also affect benthic animals. Many benthic animals have pelagic larvae and some plants have floating seeds that are transported with the water movement. Some algal blooms can be highly influenced by hydrodynamics, which accumulate the algal biomass and thus initiate or enforce a Harmful Algal Bloom (HAB) event (Sacau-Cuadrado et al., 2003; Lee and Qu, 2004). In water bodies with high flow velocity, some of the aquatic plants cannot survive while others do. Hydrodynamics can not only carry and transport algae, plant seeds or animal larvae, it can also play a vital role in nutrient transport, including the horizontal dispatch of nutrients and the vertical upwelling and turbulence mixing in aquatic ecosystems. Once the transport of water (hydrodynamics) is determined, also the transport of substances carried along with the water is determined. In return, the collective species in the water could influence the hydrodynamics, e.g. submerged rooted plants may change the hydrodynamic flow resistance due to their impact on the roughness of the bottom, flow velocity and water depth (Uittenbogaard, 2003; Baptist, 2005).

### Temperature

Many rates of change are affected by temperature, including physical characteristics (molecular diffusion), chemical (reaction) rates, and physiological rates (growth, respiration, feeding, excretion, etc.). Temperature also affects the solubility of many substances and therefore its exchange across the air-water interface. Biological reaction rates may double or even triple with every 10 degree increase in temperature and above or below critical values (Connors, 1998), certain enzymes may become inactive and organisms may die. The response of individual organisms to temperature is one where rates gradually increase towards a clearly defined temperature optimum, above which rates decline. When viewed at the ecosystem level, throughout the different seasons, ecosystems are characterised by a succession of species that become acclimated to different thermal conditions.

Temperature can not only influence the growth of populations, it also can build up stratification in the water bodies such as lakes and ocean (Wetzel, 2001). Because of stratification, primary producers, e.g. phytoplankton can remain for months in the upper, lighter layers of lakes or oceans where they can grow. However, when they die and sink into deeper waters, the nutrients released by decomposition accumulate in the deep waters and are inaccessible to primary producers because of the stratification in summer. Only when vertical mixing occurs, nutrients from the deeper water may become available.

### Light

The primary producers process photosynthesis during the day, expressed by

$$\text{Light} + CO_2 + H_2O \rightarrow CH_2O + O_2 \qquad (2.1)$$

A carbon source ($CO_2$ or organic carbon) and energy source (Light or other energy) are needed for aquatic plant growth. With increasing light intensity, the relationship between light and photosynthetic production becomes different (Figure 2-3). When light intensity increases from zero, first photosynthesis increases with the increase of light intensity, then the photosynthesis is saturated in a certain range of light intensity, and finally, photosynthesis decreases at high light intensity. The light may affect species composition because different primary producers have different favourable light density. In addition, light influences their distribution in the water due to the penetration of light and the location of primary producers.

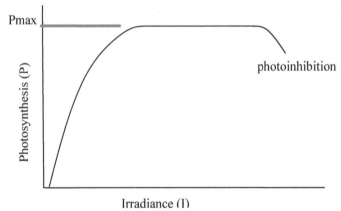

Figure 2-3 Relationship between photosynthesis and light (irradiance) (Wetzel, 2001)

Within the marine and fresh waters, the decline of solar radiation with depth effectively divides the water column into an autotrophic and heterotrophic part. This is essential for understanding the (vertical) distribution of the biotic and abiotic components (nutrients) in the water column. The photosynthesis of any kind of plant needs light, especially for submerged water plants. For submerged plants in lakes, and ponds, as long as the light can reach the bottom, plants can grow, otherwise, it is very difficult for most of the submerged plants to survive. Different species tolerate different light conditions. The amount of light that reaches sediment surface and/or the submerged plants is dependent on the transparency of water, which in turn depends on the amount of suspended organic matter (e.g. phytoplankton) and sediment particles in the water column as well as the colour of the water itself. The extinction coefficient ($k_d$) is one of the measures for vertical light attenuation, which relates to the Lambert-Beer equation expressed in Eq. 2.2.

$$I(z) = I(0) \times e^{(-k_d \times z)} \qquad (2.2)$$

or $\dfrac{I(z)}{I(0)} = e^{(-k_d \times z)}$, in which $\dfrac{I(z)}{I(0)}$ can be expressed as a percentage.

Here $k_d$ is the extinction coefficient, and $Z$ represents water depth. At the water surface $Z = 0$, $I(0)$ is the reference light intensity. While the light intensity $I(Z)$ within the water body reduces with water depth.

Furthermore, solar radiation heats up the water and the exposed sediments leading to an increase of water temperature. Moreover, seasonal, diurnal and daily variations of light intensity are also very important for studying the variation of population dynamics.

## *Nutrients*

In the growth process of primary producers like aquatic plants, nutrients and some trace elements play an important role in the formation of living organic material. For algae, the main nutrients identified are nitrogen and silicon because of their relative fraction in living tissue, and phosphorus because it so often is a limiting nutrient. Growth stops when one of the required nutrients is exhausted. The first exhausted nutrient is called "limiting nutrient", which indicates that it is this nutrient that determines and limits the extent to which growth can take place. The limiting nutrient could be nitrogen, phosphorus, silicon, iron and so on, which varies for different species as well as in different areas and different periods. For example, the main limiting factors for algae growth in marine ecosystems are nitrogen and phosphorus.

The main sources of nitrogen in surface waters are (Postma, 2003): by waste loads of organic nitrogen, ammonia and oxidized nitrogen (nitrite and nitrate) and by diffuse inflow of mainly nitrate but also of organic nitrogen due to surface run-off and seepage of ground water. In non-cultivated areas, this influx is the natural influx. Organic nitrogen is decomposed by bacteria and forms ammonia. Ammonia ($NH_4^+$) is nitrified to nitrite ($NO_2^-$), and nitrite is nitrified to nitrate ($NO_3^-$). Nitrate is the most oxidised form of nitrogen; it is taken up in aerobic environments by algae, bacteria and plants, and reduced by assimilation processes, when ammonia and $N_2$ gas are formed. Nitrate is also used as a terminal electron acceptor and is reduced to nitrite, then to

NO and $N_2O$, and finally to $N_2$ gas. This pathway is called denitrification and requires a supply of organic compounds and anaerobic conditions. Nitrate is generally the primary form of inorganic nitrogen in seawater. It is usually most abundant in winter when it is not taken up by producers, whereas during summer, it may disappear from the water column when algal blooms occur or large numbers of plants grow.

Phosphorus is often an important nutrient because it may be a limiting nutrient especially in fresh water areas. This is because phosphorus can adsorb easily to soil and form particulates. High phosphorus concentrations may boost algal growth. In seawater, it is found in living organisms or as dissolved inorganic phosphorus (DIP), or dissolved organic phosphorus (DOP). Phosphorus is usually the limiting factor for primary producers in lakes and some coastal waters, but in oceans nitrogen generally is the limiting nutrient for algae growth (Postma, 2003).

Silicate is crucially important to diatoms (a particular type of algae) growth since their external skeleton, the frustules, is made of silica. If silica concentrations are not limiting, diatoms are generally the best competitors for nitrogen, phosphate and light among the phytoplankton community (Peperzak et al., 1998). Generally, a spring diatom bloom ends due to the exhaustion of available silica (Postma, 2003).

Aquatic plant species, on the other hand, have widely different nutrient requirements and tolerances. Species that are adapted to low-nutrient conditions may be intolerant to high nutrient conditions or unable to compete with those species that thrive when nutrient concentrations are high. Different species may also require different ratios of nutrients for growth (National Science and Technology Council Committee on Environment and Natural Resources, 2000). Some species as if rooted plants mainly absorb nutrients from the bottom of lakes or ponds rather than from within the water column. There are many other abiotic factors, which are not elaborated here, that are known to influence growth, for example, pH, salinity, wind, heavy metals, etc.

### 2.4.2  Mortality, intraspecific interaction and interspecific competition

*Mortality*

Mortality can happen due to environmental changes, over-aging, crowding etc. Most of the detritus from the mortality of primary producers is recycled by mortality (about 98%) (Postma, 2003). A specific form of mortality is excretion/respiration, when the organism is not dying but uses part of its biomass for maintaining its energy level. This means that mortality and excretion/respiration are the reverse process of the creation of biomass. Respiration takes place by the organism itself. It uses oxygen and produces $CO_2$ and organic and inorganic components and energy that it uses for its maintenance. Respiration acts during the whole day, whereas the photosynthetic creation of new biomass takes place only at daytime while light is available. The day-night cycle of primary producers implies that primary producers produce biomass and oxygen at daylight and decompose biomass and consume oxygen at night, which results in large fluctuations in oxygen concentration and even can bring the night oxygen levels below acceptable limits when population outbreaks occur. Bacteria decompose the dead organic material that results from mortality. The net growth rate at equilibrium circumstances is about as high as the mortality and as high as the decomposition rate of dead organisms (Postma, 2003).

Mortalities also arise due to a change of environment, grazing by grazers and intra- or interspecific competition.

### Intraspecific interaction and interspecific competition

As long as organisms/populations live in close contact with each other, interactions will occur. These interactions may determine the distribution and abundance of species in a given system, which can be positive (e.g. mutualism), negative (e.g. competition), or neutral (e.g. neutralism), and all combinations for any of the individuals or species can be possible. These interactions are the basis of community ecology, where the interactions between the same species and different species occur simultaneously.

Intraspecific interactions include interference and resources competition (Weidenhamer, 1996). They exhibit a strong density dependency. High-density populations experience periods of mortality much sooner than do those at low densities since the populations compete for limited resources and space. These high-density populations may lead to severe population fluctuations.

For interspecific competition, several characteristics differ to intraspecific interaction: (a) individuals of different species do not all use the same resources; (b) individuals of different species do not use resources in exactly the same way; (c) interspecific competition is more likely to be asymmetrical. The competition between species includes indirect (e.g. the use of limited resources including physical space, nutrients and light, tolerance to environment etc.) and direct competition (e.g. predation, toxic release).

By consuming the same resource, a particular species can lower its availability and thus exert a negative effect on other species. Indirect competition generally does not affect the maximum biomass of a system, but becomes important to species composition. The predation of predators to prey can selectively reduce prey numbers according to their preferences, and vice versa. As the lower level in the food web, aquatic plants, especially phytoplankton, are the main food for zooplankton, some fish and water birds.

### 2.4.3  Biological growth and feedback to the external environment

The growth of aquatic plants is controlled by physical, chemical and biological processes. The basic growth leading to primary production is by photosynthesis. Although all primary producers have basically the same nutrient requirements like nitrogen, phosphorous, carbon dioxide and light, they have quite different ways to adapt and optimize their resources intake in order to be able to compete with other species and survive. Different types of aquatic plant species often have their own growth characteristics.

Here we consider phytoplankton and macrophytes as the main types of aquatic plants. Phytoplankton absorbs nutrients directly from the water through their cell wall (Brönmark and Hansson, 2005). They reproduce mainly by division of cells and they float on the water surface or subsurface. There are many different types of phytoplankton, e.g. blue-green algae (i.e. *cyanobacteria*), green algae, golden-brown algae, diatoms and *dinoflagellates*, etc. (Brönmark and Hansson, 2005). In general, there is a seasonal cycle for phytoplankton growth. Highest phytoplankton growth is normally seen in spring when there is plenty of light and nutrients. A secondary peak in phytoplankton biomass occurs in the autumn, which is somehow smaller than the spring peak. Massive growth of phytoplankton is known as algal bloom or Harmful Algal Bloom if it has harmful impacts to the ecosystem.

Macrophytes can be categorized into emergent and submerged macrophytes (Brönmark and Hansson, 2005). They reproduce by seeds, and/or by rhizomes. Many aquatic plants are able to expand rapidly by growing below ground rhizomes from which new shoots grow. Some plants propagate seeds by flow or wind which then settle down into the sediment of the water bed where they can even stay for a few years, forming seed banks, then germinate when the environment is optimal.

It is well known that the abiotic environment is creating the reference frame for aquatic populations to adapt and grow. Aquatic plants need resources, so they modify their environment and change the environment for other species. Therefore the activities of aquatic plants can affect and shape the abiotic environmental reference frame (Brönmark and Hansson, 2005).

The domination of macrophytes in a water body not only can reduce the nutrients therefore limit the growth of phytoplankton, but also can have fundamental impacts on the flow pattern in the water body. Some researchers have discovered that the massive growth of floating aquatic plants can reduce the water temperature due to the reduction of light penetration to the water body (Brönmark and Hansson, 2005).

### 2.4.4 Spatial patterns and local interactions

In general, aquatic ecosystems are heterogeneous or patchy in spatial extent. The physical, chemical, and biological characteristics can be extremely variable. Physically, ecosystems change with light levels, temperature, and water currents, etc. Chemically, they fluctuate in terms of nutrient availability, ions uptake, and contaminants presence, etc. Moreover, biologically, they vary in terms of structure and function as well as static versus dynamic variables, such as biomass, population numbers, and growth rates, and so on. There is a great deal of spatial heterogeneity in all these variables, as well as temporal variability on scales of minutes, hours, day-night cycles, seasons, decades, and even larger times.

Population patterns over space and time emerge from the interactions of adaptive individuals with each other and with their environment. Spatial patterns are the result of processes ranging from nearest neighbour interactions of individual plants to the scale of the entire water body (e.g. lake or pond) as a whole, where it may affect biodiversity and ecosystem functions (Gardner and Engelhardt, 2008). The relationships between processes such as growth, competition, mortality and spatial patterns that are observed have been a central point of plant ecology studies since long ago (Watt, 1947; Franks, 1997; Dale, 2004). The description and analysis of spatial relationships within aquatic plant species is a first step to understanding spatial patterns exhibited in nature.

Plants of one species can have a positive or negative effect on the occurrence and spatial arrangement of other species, both of the plants themselves and of a range of other organisms with which they interact, especially those species living nearby. Besides, aquatic plants have intraspecific and interspecific competitions among themselves and between different species both for food and space. Therefore, unavoidably, local interactions form one of the most important aspects for determining the spatial arrangements of plant distributions.

Spatial patterns include patch size, patch locations, patch spacing, patch density, and patch species composition. For example, the spacing between certain aquatic plants may directly influence the growth of another type of plant due to the light conditions and space to grow, which can only exist in the local area with local interactions. Furthermore, the local interactions can vary with scales. At a smaller scale, the gaps between individual plants have a more profound effect on their own growth as well as on the growth of neighbouring species, causing changes in spatial patterns. Patch density in a local area is of importance since if all else is equal, plants with the highest local density grow more slowly and experience higher mortality rates. Pacala (Pacala, 1997) introduced a 'spatial segregation hypothesis' which suggested that ecological stability is enhanced by limited dispersal and local interactions (Dale, 2004). For example, in a relatively still and shallow lake, more species of aquatic plants can adapt to such environment than in a deeper or more rapidly flowing area.

## 2.5   Summary

This chapter briefly described some of the main processes and factors involved in aquatic ecosystem dynamics, including abiotic factors such as meteorological conditions, water quality and water motion, as well as biological processes and factors including biological growth and species competition. Moreover, spatial pattern developments from local interactions are introduced as a very important aspect of aquatic ecosystem dynamics. Spatial pattern dynamics can be seen as the resulting dynamics of the interactions among species and between populations and their environment.

Aquatic plant growth is simultaneously influenced by numerous factors and processes rather than only one, but in reality, somehow it is difficult to analyze such complex phenomena. One can gain some understanding of mechanisms from in situ measurements and laboratory experiments. However, mostly in situ measurements are point measurements that often do not capture the spatial dynamics. On the other hand, laboratory experiments are generally done by varying one parameter at a time and then monitor the population response. A combination of measurements, experiments, scientific domain knowledge, as well as advanced modelling simulation may be the way to achieve a better understanding of the emerging patterns exhibit in reality, as well as for predicting population outbreaks.

# Chapter 3

# Developments in modelling paradigms

## 3.1  Introduction

In aquatic spatial population dynamics, the processes may be affected by multiple factors and the factors may change due to multiple processes. In order to understand the interactions between factors and processes, not only proper data analysis approaches are needed, but also suitable modelling paradigms need to be explored. However, modelling is rather complex and challenging because of the wide range of space and time scales involved, the often limited understanding of the nature and limitations of measurement data for calibration, as well as the availability of appropriate modelling tools.

In general, there are several steps in the process of model development: (1) understanding the system one would like to model and formulating the conceptual framework; (2) identifying essential variables and processes, as well as identifying simplifying assumptions; (3) choosing appropriate modelling tools and adequate level of detail for the dominant scales, state variables, processes and parameters; (4) designing and developing the model; (5) model calibration and verification; and finally, (6) analyzing results and interpreting its practical implications.

Among the above steps, step 2 was briefly introduced in the previous chapter, while this chapter describes the development of available modelling tools in aquatic population dynamics simulation. There are many different modelling techniques available for either modelling different processes involved in aquatic ecosystems or for solving the same problems but not necessarily giving the same results. Therefore, this chapter first reviews the development of different modelling paradigms in section 3.2, and afterwards introduces the development of some specific modelling approaches including their concepts, development and applications: section 3.3 for equation-based/ physically-based models, section 3.4 for statistical and data-driven techniques, section 3.5 for discrete cellular automata type models and section 3.6 for agent-based modelling and multi-agent systems. Finally, in section 3.7, the emerging research needs are summarized based on the modelling descriptions in previous sections and the problem characteristics introduced in the previous chapters, followed by the summary of this chapter in section 3.8.

## 3.2 Overview

In a real aquatic ecosystem, many components interact in very complex ways and it is almost impossible to separate and examine processes individually (Jørgensen and Bendoricchio, 2001). The complexity and the irreducibility of ecosystems lead to the impossibility of capturing all its details. Still, by simplification, modelling tools can be helpful for achieving better understanding of ecosystems by capturing the basic characteristics of such system.

Models are simplified representation of the real world (Mynett, 2002; Price, 2006). Normally a modelling system is composed of coupled state variables, each describing a continuous quantity and evolving smoothly in a time continuum, such as population density or averaged biomass. They normally contain the forcing functions, state variables, mathematical equations, parameters, universal constants, etc. (Jørgensen and Bendoricchio, 2001). On the one hand, models can only be useful when they capture the essential features of system. On the other hand, they need to contain those features that are of interest for solving the problem at hand. For complex ecosystems, it is difficult to capture the main features due to the often limited understanding of the mechanisms involved, and the even more often extremely limited data and measurements that are available. Still, many workable models have been developed in the past a few decades due to the rapid development of computer technology, the awareness of the importance of ecosystem functions and the increased understanding of ecosystem dynamics.

This section briefly presents some history about aquatic ecosystem modelling, which follows what was reviewed in other literature related to social sciences (Troitzsch, 1997), ecological modelling (Jørgensen and Bendoricchio, 2001; Ratz et al., 2007; Jørgensen, 2008), plant population modelling (Jeltsch et al., 2008), modelling tools for aquatic ecosystems (Pereira et al., 2006) and others. Figure 3-1 shows the development of models and theories related to aquatic ecosystem models summarized based on above literatures.

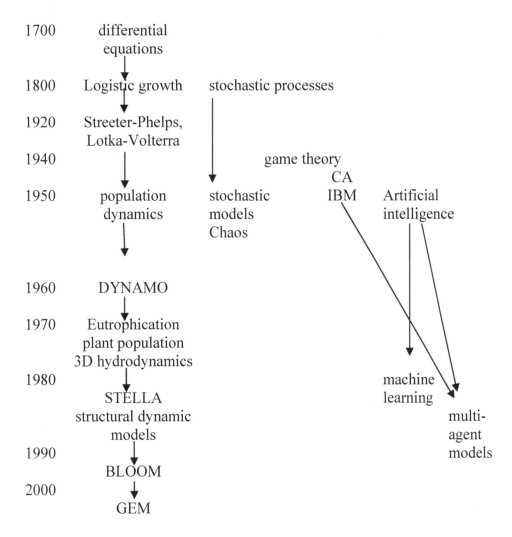

Figure 3-1 Development of models and theories related to aquatic
population dynamics modelling

Differential equations started to appear since the 18<sup>th</sup> century, and have been
widely used in many fields of science for more than 300 years. In aquatic
ecosystem modelling, the development from simple logistic growth
functions to three dimensional advection-diffusion-reaction equations, took
about two centuries. In these two centuries, not only models based on
deterministic equations were developed, also other types of models appeared
and advanced, e.g. stochastic models, chaos theory, game theory, discrete
modelling paradigms, as well as the newly emerging field of artificial
intelligence. The development of these modelling tools can be seen in
parallel with human beings' increased understanding of the natural world.
Therefore, more and more complexity has been added into the modelling of
real systems.  Since Lotka and Volterra developed their first population

dynamics model in the 1920s, which is still widely used today, population dynamics modelling has been widely developed, tested and analyzed. Table 3-1 summarizes some of the modelling approaches used and lists a few references related to the origin, reviews, and/or typical applications of these approaches.

Table 3-1 Summary of general concept, reviews and publications of different modelling approaches in ecological modelling

| Approaches | General concept, early publications, reviews and applications |
|---|---|
| Lotka-Volterra | (Lotka, 1925; Volterra, 1931) |
| Differential equations in water quality and ecology | (Holmes et al., 1994) (Postma, 2003; Postma, 2007; Los et al., 2008) |
| Population dynamics | (Levins, 1966; Levin, 1976) (Jørgensen, 2008) |
| plant population | (Jeltsch et al., 2008) |
| Eutrophication model | (Di Toro et al., 1971; Vollenweider, 1975) (Hakanson, 1999) |
| Aquatic ecosystem: | (Pereira et al., 2006) |
| Pattern and scale in ecology | (Legendre and Fortin, 1989; Levin, 1992) (Ratz et al., 2007) |
| metapopulation and spatial explicit | (Levins, 1970; Hanski and Simberloff, 1997) (Jeltsch and Moloney, 2002) |
| Chaos | (May and Oster, 1976) |
| Statistical approach | (Connor and McCoy, 1979) |
| Data-driven approach | (Recknagel et al., 2002) (Solomatine et al., 2008) |
| Spatial pattern dynamics | (May, 1993; Huston, 1994; Tilman and Kareiva, 1997) (Yokozawa et al., 1998; Wiegand et al., 2003) |
| Cellular Automata | (Von Neumann, 1949; Wolfram, 2002) (Mynett and Chen, 2004) |
| CA+vegetation/plant | (Czaran and Bartha, 1992; Balzter et al., 1998) (Dale, 2004) |
| CA+ecology | (Hogeweg, 1988) |
| CA+GIS | (Wagner, 1997) |
| Individual/agent-based in ecology | (Grimm, 1999) (Grimm and Railsback, 2005; Grimm et al., 2005; Berger et al., 2008) |

Early modelling approaches were mainly based on (differential) equations, with analytical solutions, representing equilibrium and averaged systems.

In recent decades, more population models were developed, for example deterministic three dimensional complex models, chaotic and stochastic models, static and dynamic models, distributed and lumped models, causal and black-box models, continuous and discrete models, etc. Particularly, models including spatial information and spatial pattern dynamics have become increasingly popular recently due to the simultaneous increase in both massive spatial data availability from e.g. sensor networks and remote sensing technologies together with the rapid improvements in computational power and advances in software algorithm development.

Hence, the selection of a particular model type is of importance and very much depends on the characteristics of the system, available data, user requirements and preferred modelling approach, etc. Most problems in ecology are dealing with the changes of system states with time and space. Therefore, models that are dynamics, and/or distributed, may represent these systems better since they account for the variations of variables in time and/or space.

Ratz (Ratz et al., 2007) summarized three main modelling paradigms that could be used to represent spatial pattern dynamics: (i) System Dynamics paradigm based on differential equations, either Ordinary Differential Equations (ODEs) or Partial Differential Equations (PDEs), and compartment models; (ii) Discrete paradigm including Cellular Automata (CA) and Discrete Event Specification systems (DEVS); (iii) Agent based paradigm includes Individual-Based Models (IBM) and Multi-Agent Systems (MAS). Besides the above mentioned modelling paradigms and approaches, machine learning techniques (data-driven modelling) are widely used in the last decades, especially in temporal dynamics and short term predictions (Solomatine et al., 2008). Among all these different modelling approaches, differential equation-based models, data-driven models, cellular automata as well as agent-based models and multi-agent systems are used in this research, which are briefly described in the following sections of this chapter.

## 3.3  Physically- based modelling

From the first mathematical representation of one single species to three or four dimensional numerical simulation, physically-based modelling approaches play very important role in describing aquatic ecosystem dynamics. As one of the most commonly used mathematical models for population dynamics, physically-based models are often deterministic, representing equilibrium ecosystems with averaged parameters. Such kind of modelling is based on the understanding of the underlying processes in the system one wants to model. In this case, it is a mathematical description (a set of differential equations) of how the environmental factors and their variability affect the population dynamics processes and the equations are solved numerically. Besides, differential equations arise in many areas of science and technology whenever a deterministic relationship involving some continuous quantities (modelled by functions) and their rates of change (expressed as (partial) derivatives) is known. There are mathematical models representing the basic physical and/or biochemical processes by differential equations at different levels of detail. Differential equations are mathematically studied from several different perspectives, mostly concerned with the properties of their solutions in terms of functions that represent the true phenomena as good as possible.

In aquatic ecosystems, the exhibited spatial and temporal patterns of aquatic populations are influenced both by water related properties and by ecological properties. Therefore, modelling population patterns includes processes of water motions, i.e. advection and diffusion processes, as well as chemical, biological and ecological processes, i.e. water quality and ecological processes. The advection-diffusion processes are well presented by differential equations which follow the classical conservation laws of mass, moment, and energy (Abbott, 1979). However, in water quality, and especially in ecological processes, mathematical formulations are not (yet) as well developed as the advection-diffusion equations, due to the high complexities involved in the processes themselves and the interaction among the different processes and factors.

The basic types of population models in aquatic ecosystem modelling include the modelling of age distribution, growth, and species interaction, while some models also consider the spatial extension of populations. The concepts and mathematical representations of such modelling methods are described by (Sharov, 1992) and (Jorgensen, 2001). In the past one or two decades, many researchers and research centres have developed various differential equation-based models (or physically-based models) for aquatic ecosystems and some of them considered both spatial and temporal variations for algal dynamics or other aquatic population dynamics such as macrophytes' growth, of which some examples are listed here.

One of the major concerns in aquatic ecosystems is the outbreak of populations, e.g. harmful algal blooms that can appear rapidly and sometimes are quite harmful for the whole ecosystem in the water body. Many researchers have developed physically-based models for algae population dynamics in both space and time by using conservation principles and mass-balance equations, for example, (Lee and Qu, 2004; Robson and Hamilton 2004; Romero et al., 2004; Edelvang et al., 2005; Trancoso et al., 2005; Postma, 2007; Los et al., 2008; Los, 2009). Lee and Qu (Lee and Qu, 2004) have investigated algae bloom events in the waters of Hong Kong and they developed a macro-scope HAB forecasting system based on the Delft3D hydrodynamics model for tracking the massive spring 1998 red tide event in Hong Kong waters.

On another occasion, a three-dimensional, coupled hydrodynamic–ecological model, ELCOM-CAEDYM, was applied to the period of development and subsequent decline of algae blooms in the Swan River estuary, Western Australia, by Robson & Hamilton (Robson and Hamilton 2004), who described how the model captures the complex interactions between the physical and biogeochemical environment, and indicated the usefulness of the model as a predictive management tool.

Phytoplankton dynamics is not only influenced by external forcing, but also by species competition and prey-predator processes. Some models either considered other species in the food web of the ecosystem (e.g. Armstrong, 2003) or several primary producers (e.g. (Trancoso et al., 2005). The Delft3D software system developed by Deltares | Delft Hydraulics, is a modular software package that has many different modules including hydrodynamics, water quality and ecological processes. In terms of ecological processes, the module BLOOM/GEM considers different algal species, using differential equations and optimization techniques for species composition and primary production (Los et al., 2008). This module has quite a number of applications, especially in the North Sea area (Baptist, 2005; Los et al., 2008; Blauw et al., 2009; Los, 2009) by coupling ecosystem descriptions with hydrodynamic and water quality modules.

Compared to phytoplankton, much less literature can be found on the modelling of spatial pattern dynamics of macrophytes and other aquatic plants using physically-based models (Van Nes et al., 2003; Velez and Mynett, 2006). However, several researchers have developed non-spatially mathematical formulations for aquatic plant (or macrophyte) growth (Scheffer et al., 1993; Asaeda and Van Bon, 1997; Carr et al., 1997; Asaeda et al., 2000; Madsen et al., 2001). Such type of mathematical formulations can be coupled with other spatial processes for further modelling practise. Velez and Mynett (Velez and Mynett, 2006) added a non-spatial plant growth function into the Delft3D-WAQ open process library that enabled a spatial pattern dynamics simulation for emerging aquatic plants (water hyacinths) in Sonso Lagoon, Colombia. Some literature on modelling macrophyte growth considers spatial patterns as a stochastic process (Chiarello and Barrat-Segretain, 1997).

One of the difficulties in modelling aquatic plant dynamics is a large variety of species with very different characteristics that cannot be represented only as biomass. In the literature, many references on simulation models describing the growth of particular species of interest are using detailed physiological data (Van Nes et al., 2003), which seems in general more reasonable than using a generic formulation. Still, simulation models mostly do not consider spatial extensions of aquatic plant growth explicitly, which may need another type of simulation approach beside mathematical formulas, such as empirical rules or a spatially explicit modelling concept.

In general, three main steps are distinguished in the modelling process: verification, calibration and validation (Jørgensen and Bendoricchio, 2001). Verification is the process to assess the model behaviour. It verifies the model stability, consistency, correctness of following conservations principles, as well as checks if the model performs as expected or not. Calibration is a procedure to find parameters which can lead to the best fit of observed data and modelling results, mostly by trial-and-error methods. Validation is to test how well the modelling results fit the measurements that were not used for calibration. Two types of validation can be distinguished: one is *structural* validation, used to check if the cause-effect relationship reflected by the model is adequately represented; the other one is *predictive* validation which examines whether the model has reasonable predictive capabilities in accordance with observations.

Physically-based models try to describe real world processes, and most of them are deterministic. More complex real world processes lead to more complicated models that need more reliable and extensive data for model set up, and more understanding of the processes involved. Aquatic ecosystems are complex phenomena involving both deterministic processes and stochastic processes that are sometimes difficult to be represented by physically-based models.

## 3.4   Data-driven modelling

A Data-Driven Model (DDM) can be defined as a model connecting system state variables (input, internal and output variables) and built based on available data only, without a priori assumptions on the underlying mechanisms other than assuming a specific cause-effect relationship (Witten and Frank, 2005 ; Solomatine et al., 2008). Data-driven modelling is nowadays more popular based on its capability of knowledge discovery and the increasing availability of data.

There are a considerable number of references regarding the prediction of Harmful Algal Blooms (HABs) using DDM. The most popular DDM models used in HAB prediction are Artificial Neural Networks (ANNs) (Maier and Dandy, 1997; Recknagel, 2002; Lee et al., 2003; Li et al., 2006b; Li et al., 2007b). In most of the cases in the above mentioned references, ANN turns out to be quite suitable for HAB prediction and can also be used for input selection by using a network trimming process. However, the success of neural network models largely depends on the specific problem and the influence of environmental factors on algal growth for different species. For environmentally sensitive components, it is difficult to select the dominant factors for algae growth, not only for ANN, but also for other types of data-driven modelling.

Fuzzy inference systems, which combine data and expert knowledge, are also popular methods in HAB prediction (Maier and Dandy, 1997; Urbanski, 1999; Chen, 2004; Li, 2005; Blauw et al., 2006). Since Zadeh (Zadeh, 1965) published the fuzzy set theory as an extension of classical set theory, fuzzy logic has been applied successfully in many fields where the relationship between cause and effect (variables and results) are vague. With fuzzy logic techniques, fuzzy variables are used to organize knowledge that is expressed 'linguistically' into a formal analysis. Fuzzy Logic modelling can be applied to algae bloom prediction and can be robust even in case of limited availability of data. However, the main difficulties of using Fuzzy Logic modelling are in the definition of appropriate membership functions and the induction of inference rules.

Other data-driven models like decision trees and model trees (Chen, 2004) and genetic algorithms (Bobbin and Recknagel, 2001) were also used in HAB prediction.

However, almost all the papers using data-driven modelling are non-spatially explicit HAB prediction models, which cannot represent the spatial variations in population dynamics such as bloom spatial coverage that is very important for fisheries. Some of the applications considered 2-D spatial variations by combining GIS and Fuzzy rules, for example, Urbanski (1999) evaluated the environment in coastal waters, which might be one alternative method for modelling spatial population dynamics.

## 3.5 Discrete cellular automata

From the days of Von Neumann and Ulam who first proposed the concept of Cellular Automata (CA) in the 1950s (Von Neumann, 1949), to the recent book of Wolfram 'A New Kind of Science' (Wolfram, 2002), the simple structure of CA has attracted and is still attracting many researchers from various disciplines. It is being used in many different branches of science.

Belonging to the discrete paradigm, a cellular automaton (plural: cellular automata) is a discrete model studied in computability theory, mathematics, and theoretical biology. It consists of a regular lattice or irregular grid of cells, each having one of a finite number of states. The grid can be in any finite number of dimensions. Time is also discrete, and the state of a cell at time *t* is a function of the states of a finite number of neighbouring cells (called its neighbourhood) at time *t-1*. Compared to other spatially dynamic models which are based on nonlinear differential equations, CA has a simple and natural structure and expressions. The main concept of CA lies in the fact that simple local interactions can create complex global patterns.

In biology and ecology, CA has been used to model predator-prey interactions (Conway's Game of life: (Gardner, 1970); (Resnick, 1994; Mynett and Chen, 2004), to model the growth of vegetable population(e.g. (Li, et al. submitted) and (Balzter et al., 1998)) and to HAB prediction. Cellular Automata are described as either spatially explicit models (Seppelt, 2002), or self-organization models (Bak, 1996), or object-based models (Parrott, 2005). Some of the authors consider CA together with individual-based, agent-based models as to all be object-based models, and some research on this has been done by e.g., Parrot (Parrott, 2005). The object-based modelling approach has been explored as a way to depict many of the features of a complex system over a range of scales. Among all the object-based models, Cellular Automation models seem to be one of the most common applications (Seppelt, 2002).

Most of the CA models consider synchronous updating of cell states on regular grid-cells. There are also applications and research considering evolutionary CA approaches, for example by forming asynchronous cellular automata (Suzudo, 2004) and using hexagonal cellular automata to predict wildfire (Trunfio, 2004), or irregular grids for ecohydraulics simulation (Mynett et al., 2009). The evolutions in the use of CA indicate that traditional CA can be improved in order to be more suitable for modelling reality.

The data needed for setting up CA models is normally not provided by the traditional measurements of water quality parameters and biological factors. It could be obtained from a pilot area with experimental measurements by giving different forcing to different initial conditions, but this is rather difficult and expensive, as well as time consuming. To complement the lack of data, remote sensing images have been used in this thesis, in combination with CA. CA, GIS and remote sensing images are all Grid-based; this holds great potential for integration and coupling. By coupling GIS and CA, there were already some applications in urban dynamics modelling, such as (Batty et al., 1999), and some applications in the modelling of invasive species or spreading mechanisms (Cole and Albrecht, 1999). Following the work presented in this thesis, high resolution photographs were combined with CA for spatial pattern dynamics modelling of macrophytes (Li et al., 2008b).

## 3.6 Individual / agent based and multi-agent systems

Individual-based models (also called agent-based models in some fields like social sciences) are a promising modelling paradigm, which are being widely used in ecology recently, even leading to the term "individual-based ecology" (Grimm and Railsback, 2005), defined as: *"models of individual behaviour that are useful for explaining population level phenomena in specific contexts, with contexts being characterized by the biotic and abiotic environment, sometimes including the individual's own state".*

In this thesis, the term Agent-Based Model (ABM) is used instead of individual-based model since most of the time a *group* of individuals acts as an agent or a super-individual is considered as the model unit instead of real individuals due to the large amount of individuals and the price one has to pay for computational expenses.

A Multi-Agent System (MAS) is a dynamical collection of interacting agents. An agent is an autonomous discrete entity acting on its local environment and interacting with other agents in a way chosen by itself (autonomy) and based on some knowledge of its own state and of the state of its local environment (other agents and objects of different kinds). The most crucial properties of MAS systems are: (i) the locality of the interactions between the agents and their environment; (ii) the possibility for the agents to move in this environment, i.e. to change the local environment based on interactions and consequently the entire structure of the relational network; (iii) the autonomy of the agents, which is a choice governed by the local situation; (iv) the possibility of dynamic creation or destruction of the agents.

ABM and MAS can be traced back to the Von Neumann machine concept introduced in the late 1940s, which was also the origin of Cellular Automata. From Von Neumann machine to Conway's "Game of Life" (Gardner, 1970), and Reynolds's "Artificial Life" (Reynolds, 1987), to the "Sugarscape" model created by Joshua and Robert (Joshua and Robert, 1996), agent-based models combine elements of the above theories and form its own theory. Many people also relate agent-based models to complex systems and evolutionary programming in computer science. This field was not that much developed until the first widely available computer package designed for ABM-SWARM was developed by the Santa Fe Institute at the end of the 1980's and 1990s. Nowadays, ABM is booming in many fields.

A growing literature exists on the use of MAS models for natural resource use, with a particular emphasis on land use and land cover change (Parker et al., 2003). MAS models are widely used in operational research, but applications to (aquatic) ecosystems are still limited (e.g. (Carpenter et al., 1999; DeAngelis and Mooij, 2005)). Davidsson et al. (Davidsson et al., 2007) summarized the past applications of MAS and found only 2 out of 33 to be related to ecological systems. Also, Weber (Weber et al., 2006) found there are few applications of MAS in aquatic ecological modelling. Most MAS applications use bounded rationality assumptions to limit them to a few agents only, although some, for example Berger (Berger, 2001; Berger et al., 2008) combined this approach with individual optimization. The focus of MAS models tends to be on the interaction between agents themselves, but there is great, and relatively unexploited, scope to use them also to look at the relationship between agents and their environment. Hence MAS models offer a promising approach to modelling the interaction between individual users of an open-access resource, mediated by resource availability.

Individual-based models and multi-agent systems belong to the so-called agent-based modelling paradigm which focuses mainly on individual / agent behaviour. ABM can represent an agent entity, often representing a particular class of species rather than an individual element. This also holds for macrophytes whose growth, decay and spreading is collective. MAS are a collection of interacting agents that can couple different components within a model considering both local interactions and individual behaviour in a very flexible way that is more applicable in terms of a better representation of reality and practical use (Ferber, 1999). Therefore, it seems promising to use multi-agent systems for aquatic populations' (e.g. macrophytes) collective behaviour as a discrete representation in the spatial domain.

## 3.7    Emerging research needs

One of the main research needs in ecological modelling is how to describe the spatial distribution which is often crucial to understand ecosystem reactions to system changes due to natural disasters or man-made disturbances (Jørgensen and Bendoricchio, 2001). In the past, the development of spatial pattern dynamics in aquatic ecosystem modelling had some limitations related to the large amount of input data required, the difficulty of computation of large spatial arrays, the conceptual complexity involved in the large simulation programs and the understanding of the system complexities. Since the increasing availability of remote sensing data and GIS systems to store and analyze spatial data, and the development of parallel computer systems for rapid processing, the problems related to handling large input datasets and computational costs are being eroded. However, there still remain gaps in understanding system complexities and in forming biological/ecological knowledge into mathematical representations in models.

Since ecosystem behaviour varies with spatial and temporal scales, it is vital to recognize the changes taking place at different scales, the interactions between different scales, as well as the spreading of spatial patterns through different scales. However, classical theories of population ecology only include models at the population or community level, and the theory of behavioural ecology typically addresses only the individual level and not how individual behaviour explains emergent system behaviour (Grimm and Railsback, 2005).

The key to understanding how information is transferred across scales is to determine what information is preserved and what information is lost as one moves from one scale to the other (Levin, 1992), although these features are normally hard to identify. Often population spreading mechanisms are still unclear, which leads to the difficulties in using conventional modelling tools, such as physically-based models. In addition to the limitation of ecological measurement data and the extremely limited understanding of biological and ecological mechanisms, in modelling aquatic ecosystems another difficulty is how to aggregate existing biological information and knowledge into mathematical formulations. Currently there is a lack of criteria for this issue (Wiegand et al., 2003).

Traditional methods tend to over-simplify spatial pattern variability and to under-estimate process complexity due to the limited understanding of the mechanisms involved, and the use of simple state variables at a single scale. Hence alternative modelling techniques and tools, especially ones that can involve multi-scale phenomena, are needed both for prediction of phenomena and for increasing the understanding of underlying processes, by revealing the most important factors and processes leading to the spatial patterns observed in nature.

One of the fundamental Axioms in (Grimm and Railsback, 2005) is that phenomena occurring at higher levels of observation (populations, communities) emerge from the traits of individuals and characteristics of the environment that together determine how individuals interact with each other and their environment. Discrete modelling techniques like Cellular Automata (CA) have the advantage to provide insight in local interactions and spreading phenomena. Because of this, they are becoming rapidly evolving alternative tools in the modelling and understanding of spatial and temporal variability in population dynamics, which can help advancing the understanding of population spreading and outbreaks. Cellular automata and individual / agent-based models, or any combination of these two methods, can describe the local interactions amongst individuals or groups of individuals as well as with their environment. Hence, these methods are viable approaches for simulating and revealing spatial pattern dynamics of aquatic populations.

Although there are already some research studies on the use of cellular automata for population dynamics modelling, it seems difficult to find practical applications that compared their results from CA models with measurements. Much further and deeper research is needed on the use of CA for the modelling and simulation of population spreading and outbreaks.

On the other hand, scales also refer to the scale of observation, the temporal and spatial dimensions at which and over which phenomena are observed (Peterson and Parker, 1998). The scale of observation is a fundamental determinant of describing and explaining the natural world. As data availability is an all along problem in ecological modelling, remote sensing images and GIS techniques are increasingly used, but still need to be further explored when considering the population spreading and outbreaks. The use of remote sensing images in the pre-processing and post processing of computational model simulations can be found in many papers; however, it is still rare to use RS images and data-assimilation techniques during the modelling process, which might be a further research topic now that long time series of continuous images are becoming available.

As far as population systems are concerned, populations are always influenced by their environment and interactions with their environment. Furthermore, ecological modelling is the product of interdisciplinary research as well. Therefore, the integration of different modelling techniques and information sources is essential in the modelling of aquatic population dynamics. Seppelt (Seppelt, 2002) discussed the issue of integrating or coupling different modelling approaches. The integration of different modelling approaches might be a valuable topic to follow, but the suitability of modelling integration still needs considerable investigation.

## 3.8  Summary

The selection of modelling tools depends on the specific problem to solve and the particular preference and experience of different researchers. More importantly, it depends on the combination of different criteria related to the problem including the suitability of different modelling tools. This chapter aimed to list some available modelling techniques and tools which have been used in the past by researchers in the field of aquatic ecological modelling with an emphasis on spatial dynamics modelling. Different modelling techniques have different ways to account for the processes involved in aquatic ecosystems. Some references have compared different kinds of modelling approaches, such as (Seppelt, 2002; Jørgensen, 2008), who gave an overview of the various modelling methods and their advantages and disadvantages. Still, to simulate the spatial pattern dynamics exhibited by detailed observations and to select proper processes and factors which can produce similar spatial patterns at the scales of interest, a *combination* of different types of modelling tools, information sources and measurement data seems the optimal way to achieve such a goal. In that sense, modelling remains as much an art as it is a science (Mynett, 1999).

# Chapter 4

# Harmful algal bloom prediction using data-driven techniques[1]

## 4.1   Introduction

There are many physical, chemical and biological processes and meteorological, hydrological and ecological factors involved in aquatic plant growth. Of the factors aforementioned in chapter 2, some are more relevant for aquatic plant growth, others are less. The identification and analysis of the relevant factors can be very valuable for evaluating the aquatic ecosystem and for predicting future patterns of plant growth. However, due to the large number of processes and factors involved in population dynamics, it is necessary to perform a sensitivity analysis of these factors in order to select the main factors which have the greatest influence. This chapter addresses a way of selecting main factors and shows how this method can be used as a predictive tool for population dynamics. One case study is discussed on main factor selection for algal growth prediction in Western Xiamen Bay in China.

Harmful Algal Bloom (HAB) events are due to population outbreaks of algae (phytoplankton) which can lead to harmful consequences for the ecosystem and/or human beings. HABs are quite complex phenomena. Due to its highly nonlinear behaviour, it is difficult to select the vital factors needed for HAB prediction. The causes of rapid algae proliferation could be numerous. Only if there is a clear understanding of how these processes interact to cause a HAB occurrence, can reliable models be developed that can identify periods and locations potentially susceptible to bloom events. At present the data and information needed, as well as the understanding of

---

[1] Based on:
1. Li, H., Mynett, A.E., Huang, B.Q. and Hong, H.S., 2007b. Main factor selection in Harmful Algal Bloom Prediction with a case study for Western Xiamen Bay. In: L. Ren (Editor), Methodology in Hydrology. IAHS publication, 0144-7815. IAHS, Wallingford, pp. 345-351.
2. Li, H., Mynett, A.E., Huang, B.Q. and Qiuwen, C., 2006. Harmful Algal Bloom Prediction using Data-Driven modelling: a case for Western Xiamen Bay of China, The Asia Oceania Geosciences Society's 3rd Annual Meeting (AOGS 2006). AOGS, Singapore.

the mechanisms involved in HAB prediction, are still limited. Hence it is difficult to provide early warning advice.

This chapter aims to provide a way for main factor selection in HAB prediction and shows the applicability of setting up a HAB prediction model for Western Xiamen Bay in China. Data analysis, data processing and model input selection are based on statistical analysis and expert knowledge. Two types of models are setup for HAB prediction for this area: Artificial Neural Networks (ANNs) and Fuzzy Logic (FL) rule-based systems. ANNs are used for both HAB sensitivity analysis and HAB prediction. Based on the sensitivity analysis results from ANN and the expert knowledge in this research area, a Fuzzy Logic rule-based system is setup for HAB prediction, and the results from both ANN and FL models are compared.

## 4.2   Background

### 4.2.1  Harmful algal bloom events

Algae are microscopic plants living in aquatic environments. They are vital primary producers for both marine and fresh-water ecosystems. Their growth is influenced by many different processes ranging from physical, chemical to biological and ecological processes and more. Most species of algae are not harmful. However, a HAB event can occur when certain types of microscopic algae grow quickly, forming visible patches of biomass. Algal blooming is a natural phenomenon, which has occurred throughout recorded history. Causes of algal blooming are numerous and include meteorological, hydrological, hydrodynamic, biological, ecological factors and more. HABs may discolour the water, therefore sometimes are called "red tides", but they can also appear green, yellow, or brown, depending on the type of algae. The direct effects of HABs are the depletion of oxygen and blockage of sunlight that other organisms need for living. In addition, some algae release toxins that are dangerous to animals and even humans. HAB impacts can cause human illness and sometimes death from ingesting contaminated fish or shellfish, lead to mass mortalities of wild and farmed fish, cause alterations of marine food chains, as well as closure of coastal businesses.

HABs can occur in marine, estuarine, and fresh water systems, and HABs appear to be increasing along the coastlines and in the surface waters around the world (National Science and Technology Council Committee on Environment and Natural Resources, 2000). Rapid increases in the number of people living, working, and recreating in the coastal zone have increased the input of nutrients into waters, and consequently, HAB events are occurring more frequently. In addition, increases in shipping (and the inadvertent transport of non-native species of algae in ballast water) and the transport of shellfish between regions and continents may also be contributing to increasing the frequency of HAB events by introducing new HAB species to other coastal waters, or moving them to new locations. A growing human population also increases the demand for food from coastal waters in the form of wild and cultured fish and shellfish. The aquaculture industry is threatened by HAB events and may contribute to their increase. Besides, modern detection methods such as satellite monitoring can also contribute to a higher chance of discovering HAB events than before (Qi, et al., 2003).

In China, HABs increasingly occur in marine areas, because of increased pollution and better detection methods (Qi, et al., 2003). HABs have been reported from 1970s in China and the most serious HABs were the ones in 1998 (Qi, et al., 2003). A large HAB occurred from September 18 to October 15, 1998 in the Chinese Bohai Sea, which affected an area of 5000 $km^2$ and produced losses of about 500 million Chinese Yuan (around 50 million euros). The marine area of Hong Kong and Guangdong province are the waters with highest frequency of HABs in Chinese coastal waters. From 1957 to 1996, 472 HAB events were recorded in this area, which means that on average about 30 HABs occur each year (Qi, et al., 2003).

HABs are serious problems for global aquatic ecosystems, so it is vital to mitigate the impacts from HABs. One of the main methods to achieve this is to predict HABs in order to be able to take measures (e.g. supplying physical, biological or chemical countermeasures (Anderson et al., 2001)). However, the data and information, as well as the understanding of the mechanisms are often limited. It is difficult to speculate on HABs occurrence and even more difficult to predict. Due to many factors and processes involved, it is very important to analyze the data and information, to reduce the dimensions in order to set up a predictive model without losing much information.

Literature review of the past decades shows that predictions were usually done based on experience of researchers, and recently also some physically-based modelling on flow patterns and data-driven modelling on meteorological conditions. Data-Driven Model (DDM) can provide a fast solution for unclear phenomena and sometimes can give good results provided large amounts of data are available. Artificial Neural Networks were used in this research to determine the dominant factors in observed HAB events.

## 4.2.2  Case study area

Western Xiamen Bay is a semi-enclosed eutrophic bay (Qi, 2003) with a total area of 53 km², located in the South Eastern coast of China. Due to pollution, many HABs have occurred over the past decades (Qi, 2003) and mainly from April to June. The main reasons for the occurrence of HABs in Western Xiamen Bay are its semi-enclosed geography and eutrophication. Since the 1980's, the rapid development of industry and aquaculture, as well as the increase in population, which accelerate the eutrophication of this Bay, has increased the frequency of HABs. Most of the pollution loads are from the Jiulongjiang River, and the second largest source of pollution is the industry waste load. The annual average water temperature is 22°C. Nitrogen and phosphorus are the main nutrients in this area in which phosphorus was recorded as a limiting factor (Qi, 2003). Because of the subtropical climate, there is an abundant number of algal species, which combine different types of algae with different favourite water temperature. Zhang (Zhang, 1993) reported that 110 algal species have been identified in Western Xiamen Bay, of which, 93 diatoms (84.5%), 13 *dinoflagellates* (11.8%), 2 blue-green algae and 2 *Chrysophyta*. The main indicator for HABs in this study area is Chlorophyll-a (Chl-a), which has very similar behaviour as phytoplankton (Chen et al., 1993).

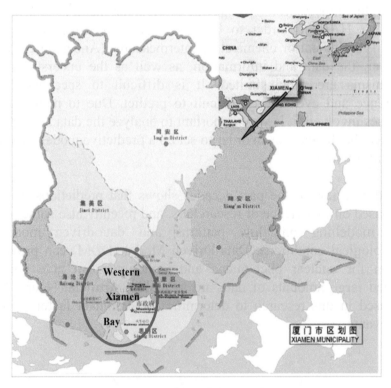

Figure 4-1 Location of the study area

Figure 4-2 Monitoring stations

In this case, data were obtained from the HAB monitoring program carried out in Xiamen in 2003. Data from 4 stations (Figure 4-2) was collected, measuring physical, chemical and biological parameters. Since historical surveys (Qi, 2003) indicated that high Chl-a concentrations can indicate algae abundance as well as the likely occurrence of HABs, this Chl-a was chosen as the indicator for predicting possible occurrence of HABs one week ahead. The ratio of total inorganic nitrogen and total inorganic phosphorous (TIN/TIP) was seen to vary from 7 to 513 while the average TIN/TIP ratio was about 33 (viz. much higher than the Redfield ratio 16), which indicates that phosphorus may have been the limiting factor for HAB events in this area in 2003.

## 4.3 Methodology

Since there are so many factors to be considered, it is very important to first analyze what data and information is available and then try to extract the dominant features in order to set up any predictive model. Conventional statistical methods can be used for multi-variable analysis and data set reduction, e.g. correlation analysis, principal component analysis (PCA), etc. Recently, also data-driven models such as Artificial Neural Networks (ANNs), have been applied for HAB prediction (Maier and Dandy, 1997; Lee et al., 2003). But in order to set up any Multi-Layer-Perceptron ANN, it is necessary to identify (i) what are the cause-effect relations and (ii) what are the parameters that dominate the process.

Clearly, no modelling – not even black-box modelling – can be done without having at least some understanding of the basic processes and mechanisms involved. Precisely for this reason, it is always advisable to start exploring any dataset using conventional statistical techniques. The results from statistical analysis can provide many hints of the relationships among the multi-factors and may lead to better understanding of the phenomena provided the statistical methods are applied appropriately.

***Correlation analysis*** is a statistical approach used to describe the degree to which one variable is linearly related to another. The coefficient of correlation $r$ is a measure of the strength of the linear relationships between two variables $x$ and $y$, and it can be computed from:

$$r = \frac{SS_{xy}}{\sqrt{SS_{xx}SS_{yy}}} \tag{4.1}$$

where

$$SS_{xy} = \sum_{i=1}^{n}(x_i - \bar{x})(y_i - \bar{y}), \; SS_{xx} = \sum_{i=1}^{n}(x_i - \bar{x})^2$$

$$SS_{yy} = \sum_{i=1}^{n}(y_i - \bar{y})^2, \; \bar{x} = \frac{1}{n}\sum_{i=1}^{n}x_i, \; \bar{y} = \frac{1}{n}\sum_{i=1}^{n}y_i \tag{4.2}$$

with $\bar{x}$ being the mean of variable $x$, $\bar{y}$ the mean of variable $y$. The closer $r$ is to 1, the stronger the linear relationship between $x$ and $y$.

***Principal Component Analysis (PCA)*** is a common statistical method for data set reduction. PCA is a way of identifying dominant patterns in the data, and expressing the data in such a way as to highlight their similarities and differences. There are some advantages in using PCA: when the data has high dimension and cannot easily be presented graphically; for compressing the data by reducing the number of dimensions without losing much information, etc. PCA has been widely applied in data reduction (Legendre and Legendre, 1998; Weiss and Indurkhya, 1998; Park and Park, 2000; Chen, 2004; Li et al., 2007b).

Data reduction using PCA can be achieved following several steps: (1) subtracting the mean of the data; (2) calculating the covariance matrix; (3) calculating the eigenvectors and eigen values of the covariance matrix; (4) leaving out some of the less significant eigenvectors, choosing principal components; (5) selecting the variables which have significant factor loading with the principal components. Those variables, which are not contributing much to the variance of the components, can be eliminated from further consideration, according to (Haan, 1977).

***Artificial Neural Networks (ANNs)*** are one of the well-established technologies in machine learning, and a mainstream technology for data-driven modelling (Solomatine et al., 2008). Data-driven modelling is nowadays quite popular because of its capability to detect trends from observed data and to provide fast predictions. It was inspired by neuroscience but did not attempt to be biologically realistic in detail. It combines simple processing elements (called neurons, units, or nodes), and the learning process in ANN is typically one of changing the strength of connections (weights) between the neurons.

*Input factors    Hidden nodes    Output*

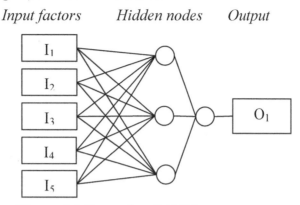

Figure 4-3 Example of ANN structure

Artificial Neural Networks (ANNs) are able to assign different weights to the input parameters and map the inputs to outputs even when the relationships between them are unclear (Murray, 1993). A common ANN configuration is the feed-forward Multi-Layer Perceptron neural network (MLP). It is made up of several interconnected nodes, arranged into three types of layers: input, hidden and output layer, with weighted connections between all nodes (Figure 4-3). The nodes in the hidden layers receive signals (values) from the nodes in the input layer and transform them into signals, which are sent to all output nodes. The transfer functions between input layer and hidden layer or between hidden layer and output layer could be linear/logarithmic or otherwise, according to the model to be built. Based on the functions selected, the model could be highly non-linear.

The training process is actually an optimisation process to minimize the errors between the outputs and actual measurements. In MLP networks, the process for solving this optimisation problem is conventionally a Back Propagation (BP) algorithm (method of steepest descent). It involves the computation of errors and the propagation of errors back through the network in order to update the weights accordingly. Practical issues for training ANNs are: establishing size and scaling factors for the data, determining the number of hidden nodes, choosing appropriate activation functions and optimisation algorithms, and prescribing performance criteria.

In this study, ANNs are used for extracting the dominant factors that determine HAB occurrence as well as for HAB prediction in Western Xiamen Bay. The software used in this study is the Weka package, which is developed at the University of Waikato in New Zealand (Witten and Frank, 2005 ).

***Fuzzy Logic (FL)*** modelling combines data and expert knowledge. This is quite useful when there are limited data available and the process is too complicated to be modelled by physically-based models. Fuzzy Logic emerged as a general form of logic that can handle the concept of partial truth. It allows easier transition between humans and computers for decision-making and a better way to handle imprecise and uncertain information (Czogala and Leski, 2000).

The process of FL modelling involves: (i) construction of membership functions; (ii) determining fuzzy logic operators; (iii) setting up if-then rules. In general, the model set-up can be described by Figure 4-4.

The difficulties of using FL are in setting up membership functions and in finding appropriate rules. Increments in the number of input factors lead to an exponentially increasing number of rules, which in turn leads to more complicated models. In this research Fuzzy C-Means clustering and expert knowledge are applied for membership function generation and inference rule induction.

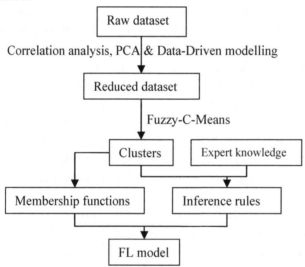

Figure 4-4 Fuzzy Logic (FL) model structure

Fuzzy C-Means (FCM) is a data clustering technique where each data point belongs to a cluster to some degree that is specified by a membership grade. This technique was originally introduced by (Bezdec, 1981). It provides a method that shows how to group data points that populate some multidimensional space, into a specific number of different clusters. Clustering techniques divide the data samples into natural groups, allowing membership functions to be formed for each separate group. The centres, the minimum and the maximum values of the clusters are used for the membership function generation. The centres of the clusters are the values, which have membership degree equal to one. Some past research has suggested that in the FL approach usually have no less than five linguistic terms (Von Altrock, 1995; Czogala and Leski, 2000; Chen, 2004), five clusters were created for Chl-a data in this research.

A membership function is a curve that defines how each point in the input space is mapped to a membership value (or degree of membership) between 0 and 1. A FL model can be described by Eq. 4.3.

$$Chl - a_{t+dt} = f(I_{1t}, I_{2t}, Chl - a_t, ...) \qquad (4.3)$$

where, $f$ is the inference rule and $dt$ is the time step in this model, $I_{1t}$ and $I_{2t}$ are the inputs for the FL model, $Chl-a_t$ are the Chl-a values at current time step, $Chl-a_{t+dt}$ represents the Chl-a values at the next time step.

In order to induce the fuzzy rules for the FL model, several steps could be followed: (a) generating empirical rules based on expert knowledge; (b) learning from available data (fuzzification) to generate the primary rules; (c) reducing the rules. Empirical knowledge is very important for setting up FL models, especially when not enough data are available. Learning from the available data means fuzzifying the numerical training data into linguistic classes, and then trying to find relationships between inputs and output.

The input for the defuzzification process is a fuzzy set (the aggregate output fuzzy set) and the output again becomes a single number. As much as fuzziness helps the rule evaluation during the intermediate steps, the final desired output for each variable can be expressed as a single number. However, the aggregate of a fuzzy set encompasses a range of output values, and so must be defuzzified in order to resolve a single output value from the set. The most popular defuzzification method is the centroid calculation, which returns the centre value of the area under the membership curve.

## 4.4   Results

### 4.4.1  Correlation analysis and principle component analysis

A cross-correlation analysis was carried out to assess whether there is a similar tendency at the four different locations. In this case, stations 1, 2, and 3 were found to have a high correlation (Table 4-1), but station 4 had quite a low correlation with the other three stations. Stronger hydrodynamics effects and different pollution sources could be the reasons for this difference. Therefore, only the data from stations 1, 2 and 3 were selected for model development and testing.

The correlation coefficients between all measured parameters and the Chl-a concentration one week ahead ($Chl-a_{t+dt}$) were generally low (Table 4-2). Rainfall ($R$) and Chl-a gave higher positive correlation with ($Chl-a_{t+dt}$).

Dissolved Oxygen (*DO*), *pH*, water temperature (*Tw*) and Chemical Oxygen Demand (*COD*) gave high correlation with Chl-a at the time of measurement (*Chl-a$_t$*). The correlations between nutrients and Chl-a are relatively high and negative as well, because of the nutrient uptake by algae.

Table 4-1 Correlation analysis between different stations (*St1*, *St2*, *St3*, *St4*)

| Chl-a | St1 | St2 | St3 | St4 | TIN | St1 | St2 | St3 | St4 |
|---|---|---|---|---|---|---|---|---|---|
| St2 | 0.85 | 1 | | | St2 | 0.86 | 1 | | |
| St3 | 0.93 | 0.66 | 1 | | St3 | 0.86 | 0.89 | 1 | |
| St4 | 0.07 | -0.10 | 0.27 | 1 | St4 | 0.17 | 0.22 | 0.12 | 1 |
| **TIP** | St1 | St2 | St3 | St4 | **Salinity** | St1 | St2 | St3 | St4 |
| St2 | 0.89 | 1 | | | St2 | 0.91 | 1 | | |
| St3 | 0.90 | 0.91 | 1 | | St3 | 0.97 | 0.92 | 1 | |
| St4 | 0.26 | 0.29 | 0.13 | 1 | St4 | 0.64 | 0.66 | 0.64 | 1 |

Table 4-2 Correlation analysis in the data of *St1*, *St2*, and *St3*

| | Tw | WaC | Tran | pH | Sal | COD | DO | TIP | Irr | TIN | R | Chl-a | Chl-a(t+dt) |
|---|---|---|---|---|---|---|---|---|---|---|---|---|---|
| Chl-a(t) | 0.43 | 0.63 | -0.62 | 0.79 | -0.31 | 0.79 | 0.91 | -0.66 | -0.12 | -0.53 | -0.08 | 1 | |
| Chl-a(t+dt) | 0.20 | 0.27 | -0.08 | 0.12 | -0.26 | 0.25 | 0.16 | -0.15 | -0.43 | 0.19 | 0.44 | 0.29 | 1 |

Note: *dt* is one week, *Tran* is transparency, *sal* is salinity and *Irr* is irradiance

Based on PCA, three components were found to give higher than average contributions, representing almost 80% of the total loading from the whole data set. The results from the factor loadings to these three main components indicate that the significant factors may include *pH*, *DO*, *Chl-a*, *R*, *TIN* and *Tw*, which is similar to the results from correlation analysis. Therefore, these factors were selected as the input variables for HAB prediction by ANN and FL. Since phosphorus was recorded as the limiting factor in this area (Qi, 2003), it was also included in the model.

### 4.4.2  Dominant factor analysis using ANN

Based on the results from correlation analysis and PCA as well as expert knowledge, 8 scenarios with different input variables were set up (Table 4-3). The time step for prediction *dt* was chosen to be one week. The data from stations 2 and 3 were selected for training because they contain all extreme values of input and output variables. Data from station 1 were selected for testing. For scenario 1 (S1), the input variables are selected based on the correlation analysis and PCA results. The input variables for other scenarios are selected by removing one of the input variables at a time. This procedure represents a network trimming process, starting from the most complicated network (S1) and reducing towards the dominant features. The resulting errors of training and testing for the 8 scenarios are shown in Table 4-4 and Table 4-5. The testing results are also shown in Figure 4-5.

Table 4-3 Scenarios in using ANN for input variable sensitivity analysis

| Scenarios | Input variables |
|---|---|
| S1 | TIN, TIP, Tw, pH, R, Chl-a, DO |
| S2 | TIN, TIP, Tw, R, Chl-a |
| S3 | TIP, Tw, R, Chl-a |
| S4 | TIN, Tw, R, Chl-a |
| S5 | TIN, TIP, R, Chl-a |
| S6 | TIN, TIP, Tw, Chl-a |
| S7 | TIN, TIP, Tw, R |
| S8 | Tw, R, Chl-a |

Table 4-4 Training errors in 8 scenarios applying MLP model using Weka

| | S1 | S2 | S3 | S4 | S5 | S6 | S7 | S8 |
|---|---|---|---|---|---|---|---|---|
| Correlation coefficient | 0.9837 | 0.9761 | 0.9525 | 0.9608 | 0.9234 | 0.873 | 0.849 | 0.9546 |
| Mean absolute error | 4.9431 | 8.472 | 9.9423 | 8.1605 | 9.6128 | 15.1644 | 9.4338 | 6.7367 |
| Root mean squared error | 6.579 | 10.386 | 12.4575 | 10.7034 | 11.3748 | 17.7621 | 14.8204 | 8.4752 |
| Relative absolute error | 28.94% | 49.61% | 58.22% | 47.78% | 56.29% | 88.79% | 55.24% | 39.45% |
| Root relative squared error | 23.71% | 37.43% | 44.90% | 38.58% | 41.00% | 64.02% | 53.42% | 30.55% |

Table 4-5 Testing errors in 8 Scenarios applying MLP model using Weka

| | S1 | S2 | S3 | S4 | S5 | S6 | S7 | S8 |
|---|---|---|---|---|---|---|---|---|
| Correlation coefficient | 0.5247 | 0.9367 | 0.8313 | 0.9413 | 0.9271 | 0.2189 | 0.8707 | 0.9328 |
| Mean absolute error | 11.023 | 11.0495 | 12.3508 | 8.4786 | 8.4742 | 20.5385 | 7.8753 | 7.032 |
| Root mean squared error | 23.25 | 13.1707 | 15.1474 | 10.1755 | 9.7253 | 31.688 | 11.2492 | 9.559 |
| Relative absolute error | 76.36% | 76.54% | 85.56% | 58.73% | 58.70% | 142.28% | 54.55% | 48.71% |
| Root relative squared error | 103% | 58.33% | 67.08% | 45.06% | 43.07% | 140.33% | 49.82% | 42.33% |

The training results of the 8 scenarios were all quite successful in capturing the peak values of Chl-a. This proves that MLP-ANNs have a high learning ability for the given training data. However, in the testing results, scenarios 1 and 6 gave very inaccurate results. Compared to scenario 2, the time delay of the occurrence of the peak value in scenario 1 is caused by pH and DO. From the correlation analysis, the highest correlation between DO, pH and Chl-a is when the time lag is 0. Algal photosynthesis uses $CO_2$ and releases oxygen, which leads to an increase of the pH and DO values. Therefore, higher pH and DO may be the consequences of fast algal growth. Scenario 6 does not consider rainfall and is seen to give inaccurate results, which shows that rainfall has a vital influence on the ANN performance in this case.

Scenarios 4, 5, 7, and 8 present relatively lower errors: this means that the input variables could be selected from any of these four scenarios. These scenarios have four or less input variables. Scenario 4 does not include TIP and nevertheless it gives a good result, which indicates that TIP may not be a limiting factor in this case. S5 does not include water temperature, and S7 does not include Chl-a at time t; this may indicate that Tw and Chl-a may have low influence in the network. It also shows the difficulties in input variable selection.

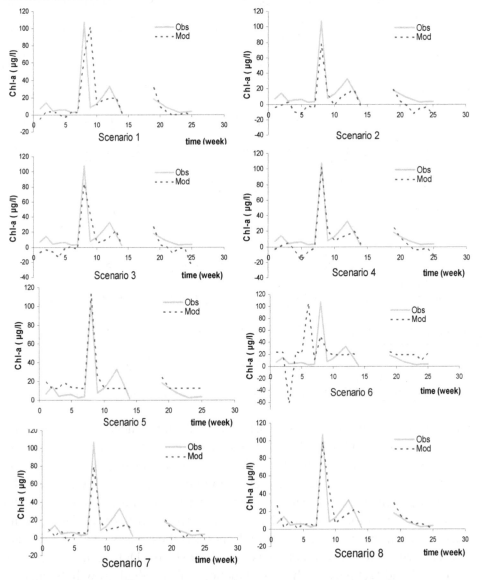

Figure 4-5 Testing results using ANN in 8 scenarios

Scenario 8 with only 3 input variables is seen to have the best performance amongst the testing of all 8 scenarios. This means on the one hand that the results do not show any advantage of using more than these three environmental factors as network inputs. It is interesting to see that such simple network already leads to quite good results. It also shows that rainfall, water temperature and Chl-a are the dominant factors for one-week ahead prediction of the Chl-a concentration in this specific case.

### 4.4.3  FL model based on the best scenario of ANN

The best scenarios from the ANN model can be considered to be scenario 8 requiring the least number of inputs and providing the best performance among all scenarios. Now, using the Fuzzy-C Means clustering method, membership functions are generated for the FL model. The membership function for Chl-a is shown in Figure 4-6.

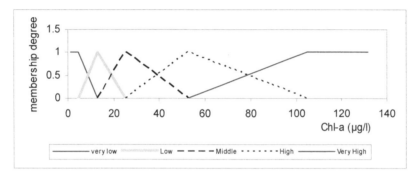

Figure 4-6 Chl-a membership functions

The inference rules are derived both from empirical knowledge and from the available data. Several rules can be generated based on empirical expert knowledge. For example:

*"if high Tw, high Rainfall and low Chl-a, then high Chl-a(t+dt)"*, or
*"if low Tw, low Rainfall, and low Chl-a,  then very low Chl-a(t+dt)"*.

Learning from the available data means fuzzifying the numerical training data to linguistic classes and then trying to find relationships between inputs and outputs, which then are to be combined with empirical rules.

The training and testing results are presented in Figure 4-7. The FL model is seen to give quite good performance for one-week-ahead prediction of the Chl-a concentration and successfully captured the peak for the main event in the testing data set based on the selected three input variables. It can be seen that statistical clustering analysis and model sensitivity analysis using ANN are quite useful for the input selection of the Fuzzy Logic model. However, compared to the results from ANN, the FL results are not really better. One obvious inaccurately predicted point is in week 22 (the end of September) in the testing result. In western Xiamen Bay, spring is the wet season and the nutrient supply comes mainly from Jiulong River and rainfall, while summer and autumn are relatively dry seasons. The inaccurately predicted point indicates that the influence from rainfall may be exaggerated for summer and autumn, or some other factors may have been important but were not considered for the summer and autumn. Accounting for the season as another input variable could be relevant for the FL model.

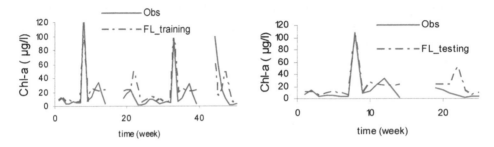

Figure 4-7 Training and testing results: FL model

One of the main advantages of using fuzzy logic theory is that it can be continuously improved by analyzing the results and adding the domain knowledge into the model, not only by learning from the data. In this case, the simple improvement of adding the season as a forth-input variable for setting up fuzzy logic model leads to training and testing results as shown in Figure 4-8 while the comparison with ANN and the original FL is shown in Table 4-6.

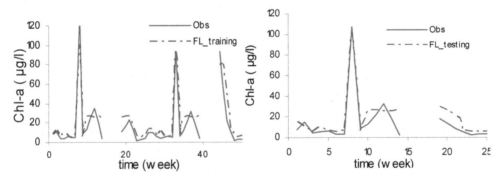

Figure 4-8 Training and testing results: improved FL model

Table 4-6 Comparison of testing errors: ANN, FL and improved FL model

|  | ANN | FL | Improved FL |
|---|---|---|---|
| Correlation coefficient | 0.9328 | 0.8642 | 0.9395 |
| Mean absolute error | 7.0320 | 8.1364 | 6.5635 |
| Root mean squared error | 9.5590 | 12.7484 | 8.9672 |

By simply adding "season" as the 4th input variable, the FL model can be improved and provide even better results than ANN. It also shows the robustness capability of using FL.

## 4.5  Summary

For the case study of Western Xiamen Bay, the ANN results showed the importance of rainfall in HAB prediction. This can be traced back to historical HABs (Chen et al., 1993). Rainfall is a very important nutrient supply carrier and indirect indicator of nutrient enrichment, especially in the period from April to June in Western Xiamen Bay. Also, rainfall may lead to salinity reduction, which may benefit the growth of some algae species.

On a more general level, data pre-processing using statistical analysis is seen to be important for not only input variability analysis and input dimension reduction, but also for achieving a better understanding of the relationship between environmental factors and HABs in this research area. ANNs already prove quite suitable tools for main factor selection for HABs and HAB prediction in this case. FL models can be (made) more robust and provide even better results, but in the Xiamen case study the differences were only marginal.

The results of this study clearly demonstrate the power of applying hydroinformatics techniques to eco-environmental data-analysis and model development; it also shows the importance of integrating hydrological processes, nutrient transport processes and algae growth processes in HAB prediction modelling for the Xiamen case study area.

In water resources management, it is important to see how aquatic plants respond to changes in hydraulic conditions and water quality management. Therefore, carrying out sensitivity analysis on multiple environmental factors affecting the growth of aquatic plants is vital. Statistical methods and models offer an alternative approach to study plant dynamics in responding to multiple environmental factors, and therefore can be used to analyze the system and predict the future behaviour of particular aquatic plant types. Worth noticing is that the selected main factors in different areas and for different species can be different due to the characteristics of a particular system and the approaches that modellers use in their analysis.

By analyzing sensitivities of multi-factors, the most important factors in a specific water body can be determined, leading to a better prediction of HAB events in time, but not in terms of spatial extension and spatial pattern dynamics. In order to predict the spatial coverage of particular harmful algal bloom events, spatially heterogeneous information and processes related to spatial dynamics need to be included in the datasets and modelling tools.

# Chapter 5

# Enhancing physically-based algal dynamics modelling using remote sensing images[2]

## 5.1   Introduction

Physically-based modelling follows from the basic understanding of the underlying mechanisms and is often represented by a set of (partial differential) equations. It is one of the main modelling approaches in population dynamics modelling. The emphasis of the model introduced in this chapter is on the spatial and temporal distributions of Harmful Algal Bloom (HAB) events. Such kind of events can lead to loss of biodiversity in aquatic ecosystems, cause serious pollution as well as mortality of many other aquatic plants and fish, and often lead to large economical losses. Therefore, it is vital for decision makers and water managers to be able to predict HAB events in order to mitigate their harmful impacts.

This chapter covers a short introduction of the research background in section 5.2, followed by section 5.3 on a more specific description of one type of physically-based modelling instruments: Delft3D-BLOOM/GEM as an example of how physically-based models are being used in the simulation of algal population dynamics. Section 5.4 describes the

---

[2] Based on:
1. Li, H., Mynett, A.E., Arias, M., Blauw, A. and Peters, S., 2008a. A Pilot study for an enhanced algal spatial pattern prediction using Remote Sensing images. In: C. Zhang and H. Tang (Editors), Proc.16th IAHR-APD Congress. Tsinghua University Press, Nanjing, China, pp. 738-743.
2. Li, H., Arias, M., Blauw, A., Peters, S. and Mynett, A.E., 2009. Enhancing Delft3D-BLOOM/GEM for algae spatial pattern analysis: model improvement, Proceedings of the International Conference on "Science and Information Technologies for Sustainable Management of Aquatic Ecosystems", the joint meeting of the 7th International Symposium on Ecohydraulics and the 8th International Conference on Hydroinformatics, Concepcion, Chile
3. Arias, M., Li, H., Blauw, A., Peters, S. and Mynett, A.E., 2009. Enhancing Delft3D-BLOOM/GEM for algae spatial pattern analysis: filling missing data in RS images, Proceedings of the International Conference on "Science and Information Technologies for Sustainable Management of Aquatic Ecosystems", the joint meeting of the 7th International Symposium on Ecohydraulics and the 8th International Conference on Hydroinformatics, Concepcion, Chile.

application of BLOOM/GEM for modelling water quality and ecological processes in the Southern North Sea area. An attempt to enhance this model with data retrieved from Remote Sensing images is described in section 5.5. A summary is provided in section 5.6.

## 5.2   Background

HABs are increasingly posing serious economic and public health problems throughout the world. The quality of many water bodies has been deteriorating over the years as a result of pollution arising from population pressure and economic growth. The occurrence of HAB events are not fully understood but are known to be vital for the balance of aquatic ecosystems. The concerns about HABs have increased over the last years mainly due to the increase in air and water pollution as well as global warming. The prediction of bloom occurrence, bloom duration and spatial extension is vital for mitigating HAB impacts. However, the occurrence of a bloom event is due to the simultaneous actions of environmental and biological factors with very complex mechanisms. In addition, the measurements which cover both spatial information and temporal dynamics are highly limited. Therefore, the prediction of HAB events, especially the spatial pattern dynamics, becomes a very difficult and challenging task.

Physically-based (PB) models are commonly used for HAB prediction and can be applied in 2 or 3 dimensions, based on the physics and bloom dynamics developing in time. PB models, in general, use lumped parameters and are often reflecting the overall system behaviour rather than the spatial pattern dynamics, due to the lack of spatially distributed measurement data. However, it is well known that the spatial heterogeneity of the flow conditions, turbidity and nutrients, which cannot be well represented by traditional point measurements at a few measurement stations, are rather important in determining the spatial patterns of algal population dynamics. Besides, PB models are in general good at capturing seasonal trends of HABs, whereas HABs are considered to be events of rapid population outbreaks characterised by temporal scales in the order of days rather than seasons. There is an immediate need for a real-time forecasting system for better short-time prediction of HAB events in order to decrease the economic impacts to coastal communities. The combination of PB and spatial monitoring with RS techniques can be of help in capturing short-term HAB events and contribute to real-time forecasting systems.

Algae blooms became of concern along the Dutch North Sea coast and Delta following the construction of the Delta Works after the 1953 flood event (Bigelow et al., 1977). Due to dense population along the Rhine-Meuse river system, inflows from nutrient loaded rivers into the relatively shallow coastal waters along the Dutch North Sea coast led to a number of algal bloom events (Los, 2009). In order to track and predict harmful algal blooms in the Southern North Sea, complex physical-geo-biochemical processes have to be taken into account. Several different models have been developed for North Sea area, which were reviewed and compared by Los et al. (Los et al., 2008), Moll & Radach (Moll and Radach, 2003) and Radach and Moll (Radach and Moll, 2006). The BLOOM/GEM North Sea application is one example of several existing European large-scale 3D ecosystem models applied to the North Sea and parts of the Northwest European Continental Shelf. BLOOM/GEM (Generic Ecological Model) is one module in the Delft3D software package developed by Deltares | Delft Hydraulics which has been used for North Sea HAB simulation for more than two decades (Los et al., 2008). The BLOOM/GEM model is capable of reproducing the conditions in the Dutch coastal waters and in general on the Southern North Sea shelf quite well. Still, there are difficulties in better representing the spatial pattern dynamics of bloom events due to the lack of spatial observations and spatial representations of forcing and/or processes in the existing PB models.

Meanwhile, Remote Sensing (RS) images are emerging as a practical and innovative alternative source of data. Remote Sensors have the advantage of collecting images in distant or difficult accessible places, such as in open oceans or at polar caps, and also near conflicting countries. In cases where directly measuring certain parameters is difficult, Remote Sensing can provide an alternative to obtain the spatial distribution of the desired parameters. Quite often, RS provides only indirect measurement information and requires detailed analyses before the results are amenable for practical implementation.

In the North Sea, turbidity as reflected by Total Suspended Matters (TSM) concentration, is an important factor influencing HAB dynamics (Blauw et al., 2009). The traditional way of modelling suspended sediment contains a high degree of uncertainty partly due to the use of steady state boundary conditions and also to river inflows and sediment loads (Gerritsen et al., 2000). Currently, the BLOOM/GEM North Sea model uses a monthly averaged suspended matter pattern derived from remote sensing data (reflecting seasonal variations) with a superimposed wind induced temporal variability, to provide model inputs for HAB simulation (Van der Woerd et al., 2005). Using this information can well represent long term amplitude variability, whereas has the danger of losing short term peaks of bloom events.

One straight forward way of adding spatial heterogeneity into (two dimensional) HAB models is to directly use the spatial data retrieved from

RS images as model input, which is one way of using RS images in this study. Unfortunately, it is uncommon to find time series of complete images due to cloudiness or insufficient satellite spatial coverage as well as distortions at the boundaries of the images. Since the GEM model requires a complete data field over the entire model domain as initial condition and input TSM data, a gap analysis is needed and missing data need to be filled in. Many spatial interpolation techniques can be used for filling in missing data in small zones, including bi-linear and nearest neighbour, cubic splines, as well as biharmonic splines (Sandwell, 1987). However, for large areas and long time missing data, these are more difficult to estimate, and one needs to combine all possible information sources in space and time, as demonstrated in this research.

As a common representative parameter for surface algal biomass (e.g. *phaeocystis*), Chlorophyll-a (Chl-a) concentration is used in this study. The Chl-a maps retrieved from RS images are as the main data for model result comparisons.

This chapter aims to enhance the HAB prediction capabilities of the BLOOM/GEM model using RS images. Spatially distributed suspended sediment data retrieved from Remotely Sensed daily images are used as the basis for BLOOM/GEM model inputs. Several techniques for filling in missing data are discussed and analyzed, in particular biharmonic splines, spatial clustering techniques and self-learning schemes for large missing data patterns. The resulting Chl-a concentration maps from the enhanced GEM model are compared with both the original GEM model and the Chl-a data retrieved from RS images.

## 5.3    BLOOM/GEM modelling instrument

### 5.3.1  Overview

BLOOM/GEM (also called GEM in (Blauw et al., 2009)) is a generic ecological modelling instrument that can be applied to any water system including fresh and coastal waters (Blauw et al., 2009; Los, 2009) for determining primary production, Chl-a concentrations and phytoplankton species composition in water bodies. This modelling instrument was originally created by WL|Delft Hydraulics, and further developed and improved over the past 20 years in cooperation of other Dutch marine research institutes: NICMM/RIKZ, NIOO-CEMO, IBN-DLO (presently Imares) and NIOZ. The goal of this generic model is to integrate the best aspects of existing models that are often dedicated to one particular ecosystem or to a subset of the relevant processes. The processes in GEM are implemented in the Delft3D-WAQ Process Library, a module for water quality and aquatic ecological modelling in the Delft3D software package (WL | Delft Hydraulics, 2007). General information about BLOOM/GEM and its main model features are summarized here, based on the description by (Los et al., 2008; Los, 2009).

BLOOM/GEM modelling instrument has been extensively used in the North Sea area for algae dynamics simulation. There are other models applied in the North Sea areas as well, for example, NORWECOM, GHER, ECOHAM and ERSEM (Moll and Radach, 2003). One of the main differences between BLOOM/GEM and these other models is that BLOOM/GEM uses a curve-linear boundary fitted grid while other models use rectangular grids. The advantages of using curve-linear grids are that they can better fit coastal lines and can easily be refined in regions where modellers are most interested in.

## 5.3.2  General description of processes in BLOOM/GEM

BLOOM/GEM includes many processes such as the transport of flow and substances, nutrient cycles, algae growth, interactions between external environmental factors with algae growth, as well as competition amongst different algae species. These processes can be grouped into two parts: (i) transport processes and (ii) water quality and ecological processes.

*Transport processes*
BLOOM/GEM calculates the transport of substances in the water column as a function of advection and dispersion induced by the hydrodynamics model, in this case Delft3D-FLOW, which is one of the modules in the Delft3D software package for modelling of hydrodynamics. The basis for computing the transport of substances can be described by Eq. 5.1:

$$\frac{\partial C}{\partial t} = D_x \frac{\partial^2 C}{\partial x^2} - u\frac{\partial C}{\partial x} + D_y \frac{\partial^2 C}{\partial y^2} - v\frac{\partial C}{\partial y} + S + P \qquad (5.1)$$

where:

| | |
|---|---|
| Dx, Dy | = dispersion coefficient ($m^2$/s) |
| C | = substance concentration |
| u,v | = components of the velocity vector (m/s) |
| S | = sources and sinks of mass due to loads and boundaries |
| P | = sources and sinks of mass due to processes |

The sources and sinks of variables on the right hand side of the equation can represent many processes in the water body, including waste loads and various water quality and ecological processes. The solution of Eq. 5.1 is achieved by using a mass-conserving scheme. Substances are either transported by the flow from one cell to the next, or converted (by a reaction equation) to other substances. This is done within each computational cell. The advection-diffusion fluxes between cells are usually derived from a hydrodynamics model. In this case, Delft3D-FLOW (WL | Delft Hydraulics, 2006a) is used to supply the hydrodynamics to the algae model. The coupling process is implemented inside the Delft3D software package. This equation is then solved numerically. In Delft3D, there are many different numerical schemes available for solving Eq. 5.1 in 1D, 2D and 3D schematizations.

### *Water quality and ecological processes*

There are many water quality and ecological processes included in BLOOM/GEM (Los, et al., 2008): phytoplankton processes (primary production, respiration and mortality); extinction of light; nutrient cycling processes; oxygen related processes, sediment-water interaction processes, etc. (Figure 5-1, State variables in grey and processes indicated by dashed lines are optional and have not been included in the North Sea modelling applications. AIP is 'adsorbed inorganic phosphorus'). Besides the above mentioned common processes included in most physically-based ecological models, a special optimization algorithm for algal species competition has been added into BLOOM/GEM (Los et al., 2008; Los, 2009).

The BLOOM/GEM modelling instrument considers three nutrient cycles: nitrogen, phosphorus and silica. The carbon cycle is partially modelled, and a mass-balance of organic carbon is made. The model assumes that the availability of inorganic carbon for uptake by algae is unlimited. Furthermore, different groups of algae can be considered: phytoplankton (e.g. diatoms, *flagellates*, *dinoflagellates*, *Phaeocystis*) or macroalgae (*Ulva* 'attached' or *Ulva* 'suspended'), oxygen, suspended detritus, and inorganic particulate matter. Light availability for phytoplankton growth is calculated based on light irradiance and extinction due to suspended sediment as well as phytoplankton and other organic matter. Different formulations are available for the characterisation of grazers, microphytobenthos, bottom sediment and sediment-water exchange. The formulations can range from simple functions (e.g. grazing) to fully dynamical processes (e.g. algae growth and mortality). Depending on the objective of the application, a decision as to which ones to include must be made a priori for each model application.

BLOOM/GEM models many different phytoplankton species using an optimization approach. Different species have different requirements on nutrients and light and they have their own biological properties as well as different preferable environments. Based on this and on the basic principle of competition, BLOOM distinguishes between three different species types: an N-type representing a species under nitrogen limitation, a P-type for phosphorus limitation and an E-type, representing the state of a species under low light conditions (Los et al., 2008). The model seeks the optimal solution for all types and within the imposed limiting factors in such a way that the total net production of phytoplankton is maximized.

A more detailed description of the processes, including model formulations and process coefficients are given in (Los and Wijsman, 2007) and (Los et al., 2008; Los, 2009).

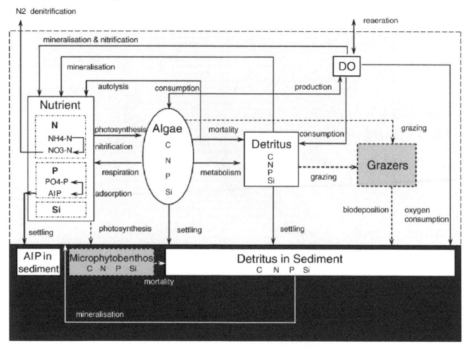

Figure 5-1 Schematic overview of all state variables and processes in BLOOM/GEM. (Los, et al. 2008)

## 5.4　Model application to the southern North Sea

The BLOOM/GEM model has been used in the North Sea area for more than 20 years, which has led to a repeated validation and continuous improvement of the model performance. The basic information of the BLOOM/GEM model for HAB prediction in the Southern North Sea area is summarised by Los et al. (2008) together with a detailed selection of characteristics. A brief summary is given here about schematization and processes included in the Southern North Sea model.

### 5.4.1  Schematization

The hydrodynamics model application introduced by (Los et al., 2008) uses the "Zuidelijke NoordZee (ZUNO) model" supplied by the Dutch Ministry of Public Works. The example in this case is a coarse ZUNO schematization (Figure 5-2). The model domain is made up of a boundary-fitted curve-linear grid, which a grid size varying from 1 km× 1 km along the Dutch coastal zone to 20 km× 20 km at the North-West corner of the model domain.

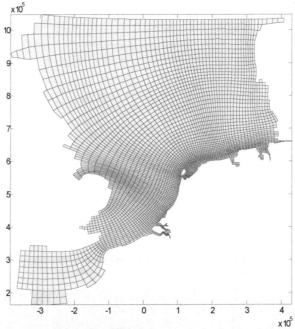

Figure 5-2 North Sea model grid used in BLOOM/GEM simulations
(known as 'coarse ZUNO grid')

### 5.4.2  Hydrodynamics and the BLOOM/GEM model

The 2D horizontal model grid of Figure 5-2 was used for allocation of Remote Sensing data. The Southern North Sea model includes the general aspects of Delft3D model configurations: coupling of hydrodynamics module to water quality and BLOOM/GEM module, providing boundary conditions and initial conditions, meteorological forcing, nutrient loads, inorganic suspended matters, etc.

Boundary conditions at the open model boundaries (astronomic tidal water levels) were obtained from simulations with a large scale model covering the entire Continental Shelf (Los et al., 2008). Coupling of the hydrodynamics and BLOOM/GEM modules as well as activating the various processes involves mainly the coupling of representative time steps for each of these processes. The meteorological data were adopted from historical measurements by the Royal Dutch Meteorological Institute, while nutrient sources were incorporated as point sources into the model domain.

One of the most difficult processes in this type of modelling concerns the suspended sediment, which directly influences the under water light climate. Therefore, the accuracy of suspended matter is very important to algae growth simulations. Much research on the modelling of sedimentation processes has been done in the past and is still going on. In this case, due to the availability of remotely sensed images and the ability of retrieving total suspended matter and Chl-a concentrations from such remotely sensed information, one straight forward way of using available information from RS images is to include the retrieved TSM as model inputs into the BLOOM/GEM model.

## 5.5   Enhanced algal spatial pattern prediction in the Delft3D-Bloom/Gem model using RS images

Remote Sensing data was made available from the MEdium Resolution Imaging Spectrometer (MERIS), which is an imaging spectrometer on board ESA's ENVISAT satellite. The atmospherically corrected data were processed using HYDROPT (Pasterkamp and van der Woerd, 2008; Van der Woerd and Pasterkamp, 2008). The retrieved TSM data and Chl-a data were calibrated based on the measurement data. The data used are daily images with 1km×1km resolution over the period from March 1 to April 30 of the year 2007.

By comparing the modelling results with the Chl-a concentrations retrieved from RS images (Figure 5-3), one can see that in general the model can capture bloom occurrences well. However, the spatial patterns and the magnitudes of Chl-a concentrations can be quite different. Among the two months of RS images explored in this study, two of the best matched days between modelled Chl-a and retrieved from RS images (April 5 and April 29) are selected and shown in Figure 5-3. It can easily be seen that the spatial distributions of Chl-a concentration shown in RS images and modelling results are quite different. One of the potential reasons could be the lack of spatially heterogeneous information supplied to the BLOOM/GEM HAB model, in addition to the complexity of the processes involved, the selection of characteristic spatial and temporal scales, and the parameters to be estimated. Therefore, an attempt was made in this research to implement TSM data retrieved from the MERIS daily data as model inputs for BLOOM/GEM.

The comparison between Chl-a data from in situ measurements on some measurement stations and Chl-a data retrieved from RS images with modelled Chl-a is given in Figure 5-4. It can be seen that the original model captures an averaging trend in these two months, but the short term peaks in Chl-a values exhibited in both in situ measurements and RS images could not be captured by the original model in this case. The variability shown in RS images can hardly be represented by the original model. Hence, further improvement on the prediction of short-term events is necessary.

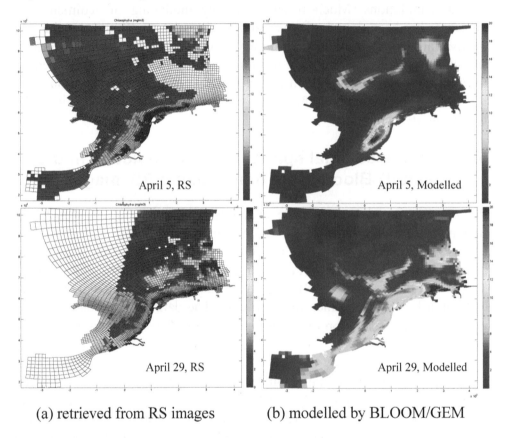

(a) retrieved from RS images          (b) modelled by BLOOM/GEM

Figure 5-3 Comparison of Chl-a concentration (mg/m$^3$) between the results retrieved from RS images and modelled by BLOOM/GEM

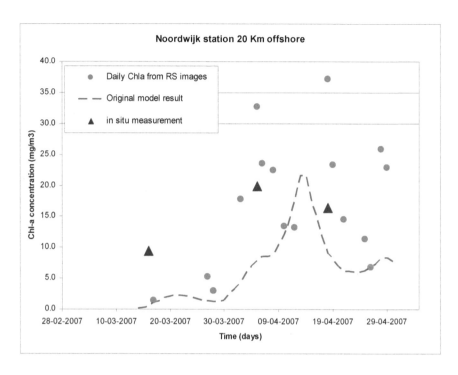

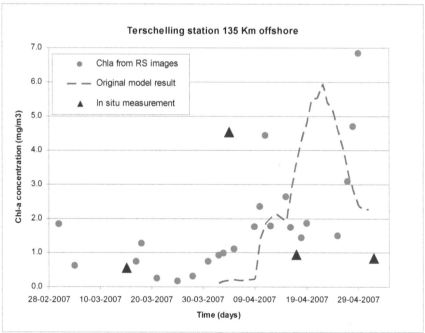

Figure 5-4 Chl-a comparison between modelled, retrieved from RS images and in situ measurement

## 5.5.1 Filling missing data in TSM maps retrieved from remote sensing images

One way of including short-term variability into the original model is to include spatially distributed information. Total Suspended Matter (TSM) is one of the most sensitive factors for HAB prediction in the North Sea. In this thesis RS images are also used to retrieve TSM maps for model inputs aside from retrieving Chl-a maps for modelling result comparison.

*Purpose*
The reasons for filling in missing data in the remote sensing images were previously discussed: not all days in March and April of 2007 had images, and even when there were images, they were generally incomplete. However, the BLOOM/GEM model requires information over the entire modelling domain. Therefore, it is necessary to construct full spatial coverage from limited available measurement data, by filling in the gaps of TSM maps retrieved from RS images.

*Approaches*
Several different spatial interpolation techniques were tried in this research, including k-nearest neighbour, kridging methods, multiple regression algorithm, bi-linear, bi-cubic and biharmonic splines, etc (Arias et al., 2009). Kridging methods (Cheng, 2000) generally require long computational time for calculations. Abdullah et al. (Abdullah and et al, 2000) proposed a multiple regression algorithm, in order to retrieve concentrations using multi band data. Many researchers also used k-nearest neighbour technique to interpolate missing data in RS images. Sandwell (Sandwell, 1987) applied biharmonic spline for the interpolation of GEOS-3 and SEASAT altimeter data. Monthly averaged spatial distributed Chl-a maps have also been used at Mediterranean Sea basin (Bricaud et al., 2002).

Considering the temporal dynamics, many techniques for time series analysis are already available (Wang, 2006), which can also be adopted to RS missing data analysis. For example, some research in remote sensing image processing considers Fuzzy Logic by applying different membership functions to different kinds of land use and soil type, mainly for agricultural purposes (Blonda, 1991). Also, in a more recent approach, tree-structured, self organizing maps (TS-SOM) were employed to estimate incomplete data (Koikkalainen and Horppu, 2007).

The optimal way of filling missing data in RS images seems to be a combination of both spatial and temporal information. Cellular Automation (CA) models commonly deal with small patches on discrete grids that consider local interactions between neighbouring cells which can lead to large-scale spatial patterns similar to ones observed in nature (Wootton, 2001). At the same time, the field of Artificial Intelligence (AI) develops rapidly, bringing more sophisticated decision rules and better interpretation and evaluation techniques. Artificial Neural Network (ANN) is one of the commonly used AI models in harmful algal bloom prediction, as discussed in previous chapter and in (Li et al., 2007b). The combination of ANN and CA holds the potential of filling large area missing data in RS images (Li et al., 2007a), which is further elaborated as Self-Learning Cellular Automata (SLCA) and is used in this study for filling missing data in large areas. A detailed explanation for SLCA can be found in chapter 6 (Section 6.2.2).

Biharmonic means that the involved function satisfies the Laplace equation two times and explicitly. This is explained by means of the use of the so-called bi-Laplacian (or biharmonic) operator in a fourth-order partial differential equation, as follows:

$$\nabla^4 F = \nabla^2(\nabla^2 F) = 0 \qquad (5.2)$$

where $\nabla$ is the nabla operator (also know as $\Delta^2$).
Based on the original idea by (Sandwell, 1987), the extended bi-harmonic equation in Cartesian coordinates (x,y,z,) (Weisstein, 2008) reads:

$$\frac{\partial^4 F}{\partial x^4} + \frac{\partial^4 F}{\partial y^4} + \frac{\partial^4 F}{\partial z^4} + 2\frac{\partial^4 F}{\partial x^2 y^2} + 2\frac{\partial^4 F}{\partial y^2 z^2} + 2\frac{\partial^4 F}{\partial x^2 z^2} = 0 \qquad (5.3)$$

Basically, the main goal of biharmonic splines is to find a functional fit (a surface, for two-dimensional RS image analysis) which passes through the available data points. Once such function is established, it can be applied to obtain functional values at all other points inside the cells. To ensure that, weights ($w_i$) or "forces" $\alpha$ are assigned to the existing cells. In order to calculate these weights, so-called Green functions $\phi$ are used. For a surface where the value of the concentration is a function of the coordinates x and y, (like in RS imagery), the particular Green function $\phi = |x|2(\ln|x|-1)$ must be used. Bi-harmonic interpolation thus implies.

$$\nabla^4 z_i = \sum_{j=1}^{N} \alpha_j \delta((x_j,y_j)-(x_{ij},y_{ij})) \rightarrow z_i = \sum_{j=1}^{N} \alpha_j \phi_m((x_j,y_j)-(x_{ij},y_{ij})) \qquad (5.4)$$

The chosen green function is evaluated for the difference between the independent values where there are available dependent values (x, y and z resp.) and the independent values whose evaluation is required ($x_i$, $y_i$ and $z_i$, in analogy). The weights $\alpha$ can then be found from the z values and the green function evaluated at the particular x and y:

$$\frac{z_{ij}}{\phi((x,y)-(x_i,y_i))} = \alpha_j = \frac{z_j}{\phi(x_j,y_j)} \tag{5.5}$$

### *Filling in missing RS data*

For filling in missing data in the TSM maps retrieved from MERIS images, many approaches are applicable but among all of the methods mentioned above (k-nearest neighbour, kridging methods, multiple regression algorithm, bi-linear, bi-cubic and biharmonic splines), biharmonic splines were selected since they seemed to give the best performance for filling missing TSM data compared to other techniques explored in this specific case study (Arias et al., 2009; Li et al., 2009a). However, it was found that almost none of the methods could give good estimates for filling in Chl-a maps retrieved from RS images. This may be due to the characteristics of HAB (rapid population outbreak) events. Chl-a spatial and temporal dynamics are a highly non-linear phenomenon resulting from the integrated effects of physical, chemical and biological processes. It changes dramatically in space and time. The Chl-a maps retrieved from RS images are in general not complete, and can only provide an indication, but cannot be used as final evidence for determining HAB events. The changes of TSM exhibited on RS images are relatively gradual compared to the changes of Chl-a on RS images.

Most of the spatially distributed TSM data retrieved from MERIS images in March and April of 2007 are with less than half coverage in the model domain. The largest variations appear near shore (viz. in the coastal zone). In the open sea, the heterogeneity is much less than in the coastal areas with its more variable bathymetry and influences from river and land inflows. Therefore, the whole domain is separated into two parts based on the water depth with threshold of 50 meters (above 50 m: deep zones, less than 50 m: shallow zones) (Figure 5-5). Potentially, a better and finer clustering within the modelling domain can be achieved by considering also hydrodynamic factors such as the characteristic flow patterns in different areas.

Based on trials with many combinations of different methods (Arias et al., 2009), scenarios could be developed for filling missing data in the specific North Sea case, using the following approach:

a) First, a visual inspection was carried out to detect potential large missing data areas with poor spatial coverage; in that case, monthly averaged images obtained from previous year(s) were preliminarily applied to fill the cells, as applied for Chl-a concentrations in the Mediterranean Sea basin by (Bricaud et al., 2002);

b) If the missing areas were not too large or were surrounded by enough cells with information, then spatial clustering was performed as shown in Figure 5-5. Bi-harmonic splines were applied for shallow zones while values from previous days (if available) or monthly averaged values from previous years (if values from previous day are not available) were used for deep-sea areas;

c) If there were two relatively good images prior to the day with a missing image, the Self Learning CA approach was used for missing data prediction;

d) Visual inspection for final check after the first three steps. If there were obvious mistakes in certain areas, those areas were replaced with values from the previous day(s) or monthly averaged data from previous years.

Table 5-1 Scenarios for filling missing data

| S | SCENARIOS |
|---|---|
| 1 | Biharmonic |
| 2 | Spatial clustering (good images: shallow: Biharmonic, deep: previous day/Monthly Average, poor images: monthly averaged) |
| 3 | SLCA based on previous two images filled by scenario 1 |
| 4 | SLCA based on previous two images filled by scenario 2 |
| 5 | SLCA for shallow zone based on previous two images filled by scenario 2 and deep zones filled by previous values |

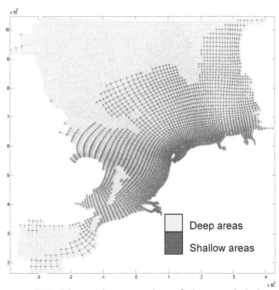

Figure 5-5 Clustering results of the model domain

### Results

The calibration of spatial interpolation methods is done by removing known values in the RS retrieved data and filling with inter/extrapolated data. The resulting errors are represented by error measurements such as root mean squared error (RMSE), mean absolute error (MAE) and linear correlation coefficient ($R^2$) comparing to the original RS data removed beforehand. More details can be found in (Arias et al., 2009; Li et al., 2009a).

As an example, resulting errors based on the above scenarios for March 26, 27 and 28 are shown in Table 5-2, while Figure 5-6 gives the original images for March 28. Figure 5-7 shows the results of some of the scenarios for March 28. The TSM maps retrieved from MERIS for the months of March and April are filled based on the calibration results.

Table 5-2 Comparison of different scenarios for filling missing data

| S | Mar. 26 | | Mar. 27 | | Mar. 28 | |
|---|---------|------|---------|------|---------|------|
|   | $R^2$ | RMSE | $R^2$ | RMSE | $R^2$ | RMSE |
| 1 | 77.2% | 3.9 | 83.6% | 2.0 | 61.1% | 1.9 |
| 2 | 84.8% | 16.2 | 93.2% | 2.2 | 82.0% | 19.7 |
| 3 | ----- | ----- | ----- | ----- | 62.7% | 2.4 |
| 4 | ----- | ----- | ----- | ----- | 86.9% | 2.5 |
| 5 | ----- | ----- | ----- | ----- | 87.0% | 2.5 |

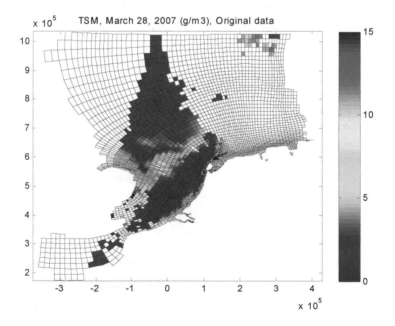

Figure 5-6 Original TSM maps for Mar. 28

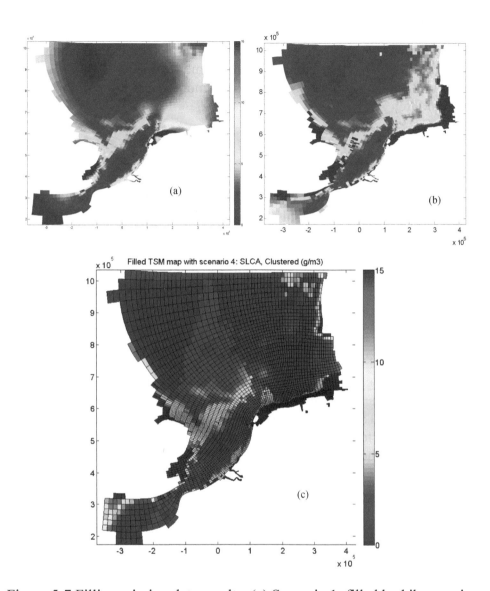

Figure 5-7 Filling missing data results: (a) Scenario 1: filled by biharmonic spline; (b) Scenario 2: filled by monthly averaged data from previous year; (c) Scenario 4: SLCA based on previous two images filled by scenario 2 (legends (a, b): same as for (c))

From the above results on using different strategies for filling missing data, one can see that different ways give different results, but most of them capture similar trends. Bi-harmonic splines give the smoothest results but with artificial distortion near the boundaries. Figure 5-7(c) shows the results with the sharpest gradients and emphasizes the high TSM concentrations near shore, which reflects the characteristics of the TSM spatial distribution in the North Sea (Los et al., 2008). In Figure 5-7 (b), the missing data on March 28 are quite large so that monthly averaged information is used to fill those gaps. Since most of the images have large areas of missing data and many of the days in the study period are completely missing, monthly averaged information becomes a rather important reference in this case. From both Table 5-2 and Figure 5-7, and some references (Los et al., 2008), scenario 4 shown in Figure 5-7 (c) seems more suitable for filling missing data in March 28.

### 5.5.2 Enhancing HAB modelling with TSM maps retrieved from RS images

When comparing the results from the original and enhanced BLOOM/GEM North Sea models with the Chl-a data retrieved from MERIS images, many daily images are seen to hardly have any similarity with the original model results in the two months of March and April, 2007 in terms of spatial pattern dynamics of Chl-a concentrations (some examples are shown in Appendix I). However, after adding TSM data retrieved from RS images as model inputs, a better representation of spatial patterns was observed in many of the Chl-a output maps. As an example, the Chl-a map from RS images and from both the original model and the enhanced model on March 28 are shown in Figure 5-8 a, b and c. The original model could not capture the high Chl-a concentration shown in the data retrieved from MERIS images. It shows the necessity of improving the model prediction results. After using the TSM data retrieved from MERIS images, one sees a much better model performance (Figure 5-8 c). The high Chl-a appearing in the RS image near the Dutch coast is simulated better and the enhanced model better reflects the spatial patterns exhibited on the Remote Sensing image maps.

Another example is shown in Figure 5-9 with a zoomed view for the Wadden Sea area on April 5. It also shows a better HAB spatial coverage represented with a high Chl-a concentration on enhanced model results (Figure 5-9 (c)) than original model (Figure 5-9 (a)) compared to Chl-a retrieved from RS images (Figure 5-9 (b)). The in situ measurement on one of the stations in Wadden Sea also showed a very high Chl-a concentration around April 5, which is consistent with the RS image and enhanced model results.

The Chl-a concentrations retrieved from Remote Sensing images (from March 1st to April 30th), and from both the original and enhanced model were tested for different measurement stations. Two examples are shown in Figure 5-10 and Figure 5-11, which represent the comparison on stations of Noordwijk 20km and Terschelling 135km respectively.

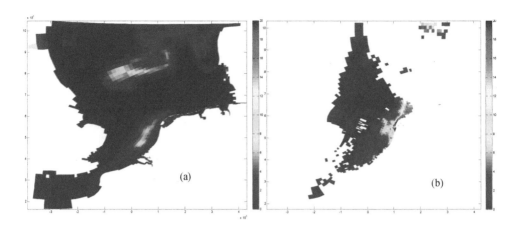

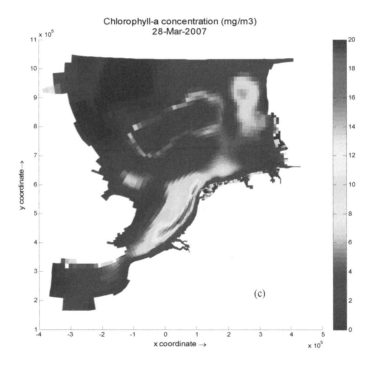

Figure 5-8 Comparison of (a) original model, (b) RS retrieved and
(c) enhanced model (Li et al., 2008a)

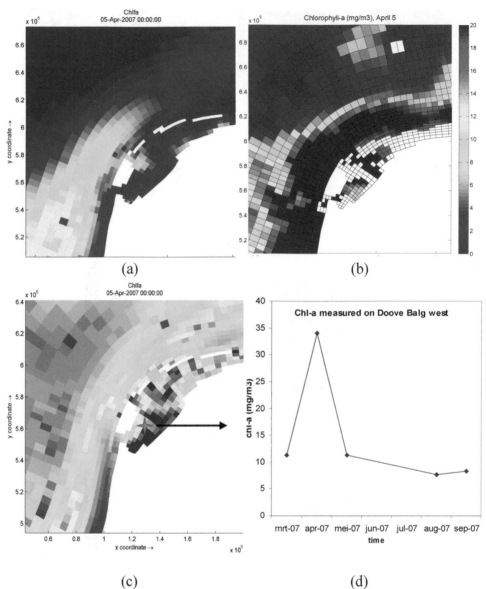

(a)                                                    (b)

(c)                                                    (d)

Figure 5-9 Comparison of Chl-a from (a) original model, (b) RS retrieved
(c) enhanced model on April 5 at Wadden Sea region, and (d) in situ
measurement in the middle of Wadden Sea (Doove Balg west station)

As seen in Figure 5-10, around March 30th in 2007 there is an increase in
Chl-a concentrations which is better simulated by the enhanced model
(scenario 5 in table 2). The enhanced model behaves well, capturing the
bloom peak from April 3rd to April 8th but behaves similar to the original
model after April 19th.

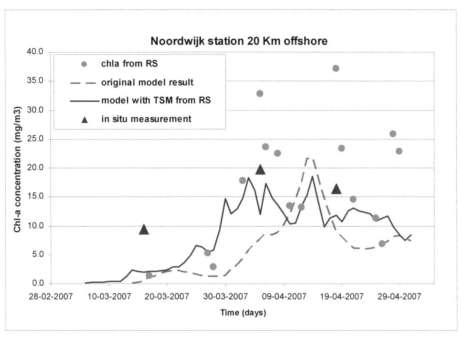

Figure 5-10 Chl-a time series on Noordwijk station, 20 Km offshore

Another example of model comparison as can be seen in Figure 5-11 (Terschelling station 135 Km offshore) shows that qualitatively the enhanced model agrees well with field data, especially during bloom peak events occurring in the first week of April, also capturing well the decreasing tendency in April, whereas the original model could not capture the trend exhibited in the measurements. The enhanced model is seen to capture better the general trend than the original model compared with the Chl-a data from remote sensing images. However, on this station, the in situ measurements do not match with the data from RS images, especially from April 26 to 29. One of the reasons is that the RS images before April 29 were not of good quality and of low coverage on the research domain. The inconsistency of in situ measurements and RS images can lead to difficulties for model calibration and comparison, as well as the reliability of model accuracy. Further uncertainty analysis is needed for both in situ measurements and the data retrieved from RS images. Several stations are with results hardly improved by the model performance, which may due to the sparse measurements, or the resolution and quality of RS images (RS images have difficulties in retrieving TSM in very shallow areas) as well as very coarse grid cells that cannot match well with the measurement station.

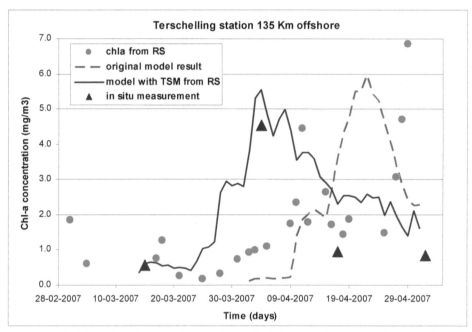

Figure 5-11 Chl-a time series on Terschelling station, 135 Km offshore

### 5.5.3  Discussions and remarks

This was a pilot study on using data retrieved from RS images as both input and data for model result comparison for the Generic Ecological Model in North Sea algal bloom prediction. It demonstrated the importance of including RS information into the conventional modelling approach and showed the advantages of using high frequency RS images for representing the large variability of phytoplankton biomass in both space and time. In addition, it also proved that the concentration of total suspended matters is a very sensitive factor in algal bloom prediction in the North Sea. With better images supplied, HAB prediction can be more enhanced. Furthermore, the results in the enhanced model still have the potential to be further improved with a better estimation of TSM by developing better techniques for filling missing data and by improving the representations of processes involved.

Filling in missing data is a very challenging task. This research tried many different ways and summarized some scenarios for the Southern North Sea case. The HAB prediction results with filling missing data method as summarized in this study are seen to be suitable for obtaining TSM maps. However, due to much higher spatial and temporal variations in Chl-a concentrations, filling missing data in Chl-a maps was seen to be not a suitable way of getting complete maps. In addition, the results of filling missing data could be improved by a better clustering in the modelling domain, (e.g. by considering variations in flow dynamics).

The GEM model considered in this study was a two dimensional model since the RS images only supply surface information rather than vertical distributions. Such 2-D models may miss some important aspects on forming HAB events: vertical mixing and subsurface algal species.

The boundary-fitted curve-linear model grids influence both the use of filling missing data techniques and model representation of the spatial heterogeneity. Some of the areas in the model domain have very similar resolution as RS images (about 1 km$^2$), but other regions are with much larger cell sizes (about 400 km$^2$) than RS images. Finer model grids may give better model results, however this may be computationally expensive.

This case study used two months of MERIS images to supply TSM inputs to HAB simulation. Therefore, the seasonal variation of HABs cannot be reflected. Clearly the approach needs to be extended to annual simulations as long as sufficient RS images (of good quality) are available in order to draw more statistically significant conclusions.

## 5.6  Summary

The Delft3D BLOOM/GEM model is based on solid mathematical formulations for hydrodynamics, water quality as well as ecological processes, and can show good model performance, especially seasonal performance in North Sea area based on the cost function calculation in (Los et al., 2008). Still, real situations are often far more complex than any model can represent and the short-term prediction of HAB events becomes a more important issue than annual and seasonal variations. Therefore further research is needed for more discovery of biological and ecological domain knowledge related to short term HAB events and the implementation of such knowledge into mathematical formulations and software systems.

The use of monthly averaged suspended matter values retrieved from remote sensing data and superimposed by wind may not reflect the short-term spatial and temporal dynamics well for Total Suspended Matter (TSM), which may be one of the reasons that the original model could not capture well the Chl-a spatial patterns on some days. Clearly these types of models could be enhanced by: (a) adding more temporal and spatial data, as demonstrated in this chapter; (b) providing a better formulation about the relationships between abiotic and biotic factors, e.g. light-climate (Los et al., 2008); and (c) by including additional processes and mechanisms which are lacking in conventional approaches, e.g. biological/ecological diffusion, stochastic aspects and discrete (burst/event) phenomena.

In this chapter, it was shown that one possible way of enhancing the BLOOM/GEM model performance is to include more spatial heterogeneous data relevant to bloom occurrence. Other developments in data-model integration for HAB simulation are likely to further enhance conventional modelling approaches, leading to even better predictions.

In this case, we only simulated spatial pattern dynamics of Chl-a concentrations. However, one of the advantages of using BLOOM/GEM is that it can represent competition between multiple species for different resources, which leads to concentrations of different species rather than only Chl-a. Nevertheless, modelling spatial pattern dynamics of different algal species requires even more spatial data on different species to be available.

# Chapter 6

# Spatial pattern dynamics in cellular automata based aquatic ecosystem modelling[3]

## 6.1 Introduction

Aquatic population growth involves a number of physical, chemical, ecological and biological processes. Modelling such complex phenomena is rather difficult and still far from being accurate. On the one hand, the lack of process understanding makes it difficult to be modelled by physically-based modelling tools, while on the other hand, limited availability of adequate spatial and temporal measurements data for water quality and ecological parameters makes the validation and calibration of any model very challenging. Cellular Automata (CA) models are capable of dealing with spatial variations and local interactions. They have become increasingly popular in ecosystem dynamics modelling since simple local rules can lead to globally complex patterns. In the meantime, remotely sensed images and high resolution photographs are becoming more available for monitoring spatial pattern development, which makes it possible to model population spatial pattern dynamics with spatially explicit methods, such as Cellular Automata. In this chapter, the concept and general description of Cellular Automata as well as several types of a cellular automata based models are addressed in section 6.2. One case study, which combines time series of high resolution photographs with meteorological data and biological knowledge, is presented in section 6.3. The applicability of Cellular Automata to capture the biological characteristics of aquatic population

[3] Based on:
1. Li, H., Mynett, A.E. and Corzo, G., 2007a. Model-based training of Artificial Neural Networks and Cellular Automata for rapid prediction of potential algae blooms, 6th International Symposium on Ecohydraulics, Christchurch, New Zealand.
2. Li, H., Mynett, A.E. and Chen, Q., 2006a. Modelling of Algal Population Dynamics using Cellular Automata and Fuzzy Rules, Proc. 7th Int. Conf. on Hydroinformatics. Research Publishing, Nice, France, pp. 1040-1047.
3. Li, H., Mynett, A.E. and Penning, E., 2009, photography-based cellular automata in aquatic plant dynamics modelling. Submitted to Journal of Ecological Informatics.
4. Li, H., 2007. Scale coupling in algal dynamics modelling using DELFT3D with Fuzzy Inference Complex Automata, Proceedings of 32nd Congress of IAHR, the International Association of Hydraulic Engineering & Research, Venice, Italy, pp. 782.

growth is discussed based on this case. A summary is given in section 6.4 which concludes this chapter.

## 6.2 Cellular automata

### 6.2.1 Background

Cellular Automata (CA) have attracted many researchers in various fields since the 1950s (Von Neumann, 1949). Currently, at least one journal on *Complex Systems* is primarily devoted to CA. Cellular Automata conferences are held every two years with proceedings and lecture notes in computer science. Many books and a huge number of publications can be found in journals about the use of Cellular Automata. The most well known books are *Cellular Automata machines: a new environment for modeling* (Toffoli and Margolus, 1987), and *A new kind of science* (Wolfram, 2002), which have very high citation rates.

CA models deal with spatial variation and local interactions. They provide simple discrete deterministic mathematical models for physical, biological and computational systems in which many simple components act together to produce complicated patterns of behaviour (Packard and Wolfram, 1985). A cellular automaton consists of a regular uniform lattice of cells, usually infinite in extent, with a discrete variable at each cell (Wolfram, 1986). A cellular automaton evolves in discrete time steps. The variables at each cell are updated simultaneously, based on the values of the variables in their neighbourhood at the preceding time step, and according to a predefined set of "local rules". The neighbourhood of a cell is typically taken to be the cell itself plus all its immediately adjacent cells. There are a great numbers of particular CA models, each one being a specific system. However, most CA models have the following five generic characteristics (Ilachinski, 2001):

- Discrete lattice of cells: the system substrate consists of a one-, two- or multi-dimensional lattice of cells.
- Homogeneity: all cells are equivalent.
- Discrete states: each cell takes on one of a finite number of possible discrete states.
- Local interactions: each cell interacts only with cells that are in its local neighbourhood.
- Discrete dynamics: at each discrete time step, each cell updates its current state according to a transition rule taking into account the states of cells in its neighbourhood.

In the case of ecosystem modelling, the use of simple rules to form global complexity by considering local interactions between populations and their environment, seem to have great potential of representing e.g. aquatic plant growth dynamics. However, finding appropriate rules for a cellular automaton model is a crucial task (Balzter et al., 1998).

It needs the most attention in finding proper spatio-temporal dependence. There are several possibilities relevant to aquatic population dynamics modelling for constructing the local rules $f$ in CA model: (a) simple if-then rules based on the data and the empirical or experimental knowledge (Li et al., 2009b); (b) weighted rules which emphasizing on the importance of distances or other factors induced biases of neighbouring cells' contribution, e.g. flow pattern; (c) Fuzzy Inference Rules based on the training of the available data and combining expert knowledge into the set of rules (Chen, 2004); (d) probabilistic rules to form a kind of stochastic CA model ; (e) rules automatically generated by the available data and using machine learning techniques (Li et al., 2007a), e.g. Artificial Neural Networks, which is more suitable for the applications with sufficient amount of time series data.

## 6.2.2 Types of CA models

Based on the types of grids and different kinds of local rules, CA models can be catalogued into many different types, among which traditional CA, stochastic CA, self-learning CA and Fuzzy-rule based CA are described in the following paragraphs.

***Traditional Cellular Automata*** are based on a domain of discrete regular grids. The local rules are simple if-then rules deduced from expert knowledge. Conway's Game of Life is a typical example of such kind of Cellular Automata. It considers very simple transition rules but can produce many different patterns (Gardner, 1970).

A simple CA can be defined by a lattice $L$, a state space $Q$, a neighbourhood scheme $N$ and a local transition function $f$ (Balzter et al., 1998):

$$CA = <L, Q, N, f>$$ (6.1)

The lattice $L$ can be triangular, square, hexagonal, etc. The most commonly used lattice is a square lattice. The discrete state for $L$ comes out of $Q$.

Two-dimensional CA can be defined as a double array of lattice and can use many different neighbourhood schemes. For a square lattice, the classical neighbourhood schemes include (i) a five-cell scheme (Von Neumann), (ii) a nine-cell scheme (Moore), and even (iii) a thirteen-cell scheme (Extended Moore), etc (Figure 6-1). Classical neighbourhood schemes consider equal weights for all the neighbouring cells, and the rules are applied homogeneously in the whole domain.

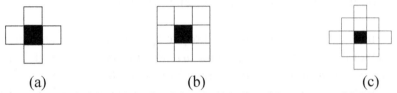

(a)                              (b)                              (c)

Figure 6-1 Neighbourhood schemes: (a) Von Neumann, (b) Moore, and (c) Extended Moore

The design of the CA model considers not only neighbourhood schemes, but also space and time steps, initial conditions, boundary conditions and inference rules. The initial conditions can be from RS images, photographs, or the results from other models. The grid size and time step selection depends on the growth properties of the considered plant and the model's computational capabilities.

The cells change their states $S$ at discrete time-steps based on the current states of the neighbouring cells through certain transition rules $f$ (Eq. 6.2).

$$S_i^{n+1} = f\,(S_i^n,\ S_{i1}^n,\ S_{i2}^n,\ S_{i3}^n,\ S_{i4}^n......)  \tag{6.2}$$

where: $n$ is the current time level, $n+1$ is one time step ahead, $S_{i1}$, $S_{i2}$, $S_{i3}$, $S_{i4}$ etc. are the neighbouring cells of $S_i$, while each cell uses the same rules $f$.

The initial pattern constitutes the '*seeds*' of the system. Each generation is a function of the previous generation and the rules are applied simultaneously to every cell in the whole domain at each discrete moment in time. The rules continue to be applied repeatedly to create a next generation and so on.

Boundary conditions need to be defined for the research domain. The selection of boundary conditions is mainly based on the particular research problem. If the research domain is a closed water body like a pond or lake, fixed/closed boundary conditions can be applied. However, if the research domain e.g. has a mass exchange across the boundaries (like in the case of a coastal zone), open boundary conditions have to be supplied.

***Stochastic Cellular Automata*** are particular types of CA with a probabilistic behaviour, which is not like most cellular automata, whose behaviour is deterministic. Stochastic cellular automata are models of "noisy" systems in which processes do not function exactly as expected, like most processes found in natural systems. By adding some noise in the transition rules or by deriving probabilistic transition rules, we can develop a stochastic CA. It is not always necessary to be based on a domain with regular grids. Minns and Mynett (Minns et al., 2000) used a stochastic CA to simulate prey-predator systems and compared it with the classical Lotka-Volterra (LV) model. They observed a similar trend but noted that stochastic CA included more detailed spatial information (patchiness) that was not contained in the classical LV model, due to its assumption of homogeneity and averaged parameter representation. In Stochastic CA, more noise can be included and chaotic behaviour can occur more easily.

The behaviour of these cellular automata tends to be very rich and complex, often forming complex spatial patterns. They are capable of mimicking many phenomena found in nature such as crystal growth, boiling of fluids, and turbulent flow phenomena. However, a major problem with stochastic population models is the influence of unobserved (and often unobservable) variables that affect the probability structure over time (Balzter et al., 1998). Exhaustive sampling of all relevant factors can be too expensive to be implemented. Besides, stochasticity in reality is not always measurable.

***Self-Learning Cellular Automata (SLCA)*** is defined here as a Cellular Automata model develops its own transition rules using data-driven techniques like Artificial Neural Networks.

Artificial Neural Networks, especially of the Multi-Layer Perceptron (MLP) type, are successfully being used to represent many nonlinear phenomena. They tend to be able to capture a number of variables in one unique representation that do not allow for intermediate interpretation (i.e. space and time variables). Therefore, different types of data-driven models need to be developed to focus on spatial dynamics that consider states of interaction.

For complex phenomena with limited mechanisms of understanding, traditional transition rules such as if-then rules sometimes are insufficient to represent non-linear relations among neighbouring cells. The use of ANN could possibly supply an automatically learning mechanism for such non-linear transition rules (Figure 6-2). Therefore, the combination of artificial neural networks and CA may be a useful alternative modelling method to supply rapid prediction with the consideration of spatial interactions. Furthermore, such combined method can be useful to identify the most sensitive variables in the model representation, as an extension of an earlier application in Chapter 4.

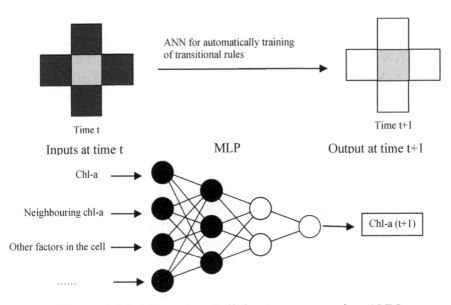

Figure 6-2 Self-learning Cellular Automata using ANN

Such kind of combination of ANN and CA has previously been used in the modelling of urban development (Almeida et al., 2008) and in the emulation of physically-based water quality modelling results (Li et al., 2007a).

In (Li et al., 2007a), a simplified North Sea Dutch Coast HAB model was setup by using Delft3D-WAQ. Modelling results were emulated by a SLCA model for predicting Chl-a concentration. As all other data driven models, the pre-processing and training details are decisive for the quality of the model outcome. The pre-processing used in this research involved a correlation analysis of the variables for a lag-limited time step in the input selection. The average mutual information between different lags and variables was evaluated to obtain a measure of the non-linear correlations. A normalization based on standard deviation was applied to all the inputs.

The learning of the MLP model is done through the minimization of the root mean square error. This minimization is based on the Levenberg Marquardt algorithm (Levenberg, 1944; Marquardt, 1963). Since this algorithm has shown fast convergence, the selected epoch number was 150 for the entire experiment. The learning rate (0.1) and momentum term (0.7) were the same for all scenarios. To select the most adequate ANN-MLP, a cross-validation technique was included based on a 10-fold sampling of training and verification data. Each fold commonly contains 90% of the data for training and 10% for verification. As a result, the best model in this process is selected.

In the paper by (Li et al., 2007a), the results in considering the values of Chl-a in the centre cell and its 4 neighbouring cells for one day ahead Chl-a prediction, were seen to give the highest accuracy compared to other model scenarios which considered either external forcing and nutrients (irradiance, suspended sediment, inorganic nitrogen and phosphorous Chl a) or only the Chl-a concentration in the cell itself. The results from Li et al. (Li et al., 2007a) indicate that the local interactions are very important for spatio-temporal algal population dynamics prediction. They show the possibility of using ANN and CA not only for future prediction of one single cell, but also for considering spatial interactions within the whole domain.

SLCA not only can be used for the emulation of physically-based models, but more importantly, can also be used to help in finding the proper spatio-temporal dependencies by varying the modelling grid size and time steps. Different spatial and temporal scales may require different neighbourhood schemes in the use of CA.

***Fuzzy-rule based cellular automata*** model rules are derived by using Fuzzy inference rules which combine both expert knowledge and data.

Fuzzy Logic (FL) modelling is quite useful when there are limited data available and the process is too complicated to be modelled by physically-based models. Fuzzy Logic emerged as a general form of logic that can handle the concept of partial truth. It allows easier transition between human interpretation and computer calculation in decision-making, and a better way to handle imprecise and uncertain information.

Fuzzy Logic modelling can be applied to algae bloom prediction and was found to be quite robust in case of limited data availability (Li et al., 2006b). Raw data can be analyzed by some statistical methods and data-driven techniques (Li et al., 2007b). Cellular Automata using Fuzzy Logic for transition rules can be described by Eq. 6.3.

$$X_{t+\Delta t} = f(X_{t(i,j)}, X_{t(i-1,j)}, X_{t(i,j-1)}, X_{t(i+1,j)}, X_{t(i,j+1)}, P_{1t}, P_{2t}, P_{3t} \ldots\ldots) \quad (6.3)$$

where, $f$ is the inference function and $\Delta t$ is model time step. $X_{t(i-1,j)}, X_{t(i,j-1)}, X_{t(i+1,j)}, X_{t(i,j+1)}$ are values of the nearest neighbouring cells of $X_{t(i,j)}$, and $P_{1t}, P_{2t}, P_{3t}$ are other factors which are included as model inputs.

In aquatic ecological modelling, data is often limited and so is the understanding of the mechanisms involved. Besides, we may understand certain mechanisms but it is hard to be implemented into models. Fuzzy Logic is one way of translating our understanding into models in combination with partially available data. Therefore, the combination of Fuzzy Logic and Cellular Automata seems an interesting alternative for spatial pattern dynamics in aquatic ecosystem modelling.

There are many other ways to construct CA-based models. For example, unstructured Cellular Automata (Mynett et al., 2009), asynchronous Cellular Automata that do not update the cell states synchronously (Schönfisch and de Roos, 1999), deterministic Cellular Automata that uses equations instead of rules as transition functions (Chen et al., 2002). Even earlier, some special kinds of Cellular Automata were introduced, such as Lattice Gas Cellular Automata for modelling hydrodynamics (Rothman and Zaleski, 1997). The selection of CA models depends on many factors of research, data availability, mechanism understanding, system properties, and user requirement etc. It is a challenging and ongoing field of research.

## 6.3  Photography-based cellular automata in aquatic plant dynamics modelling

### 6.3.1  Introduction

The spatial patterns of aquatic plants are a crucial aspect for plant growth as well as for other species in the same ecosystem. Locally, different plants compete mainly with their neighbours, while on a larger scale patterns arise that are characterized by their overall density and shape, commonly referred to as *patchiness*. Over the last decades, there is an increasing trend in research on aquatic population spatial pattern dynamics modelling (Mynett and Chen, 2004) with special attention being paid to spatial plant evolution (Chen et al., 2002; Freckleton and Watkinson, 2002; Giusti and Marsili-Libelli, 2006; Hogeweg, 2007). Among many of the modelling approaches, cell-based locally interactive models, e.g. Cellular Automata (CA), seem to have great potential for representing the spatial patterns evolving from local interactions.

In this research, a Cellular Automata model was developed for mimicking macrophytes growth in a small pond by combining time series of high-resolution photos, meteorological data and plant properties. Furthermore, this study analyzes the effect of different cell sizes and neighbourhood schemes for capturing the specific biological characteristics when setting up a CA model.

## 6.3.2 Influencing factors for water lily growth

In the food web of an ecosystem, macrophytes are sources of food and shelter for other species like fish and ducks. They contribute to nutrient cycling, flow condition stabilization, and they can also change the hydraulic roughness of the bottom of the lakes or ponds.

One particular type of macrophytes, the water lily (*Nymphaeaceae*), is rooted in the soil in bodies of shallow water (optimal depth less than about 1 meter), with leaves and flowers floating on the water surface, which is demonstrated in Figure 6-3. They are in the *Nymphaeaceae* family with round and large leaves (diameter 6-11 cm) that are able to store sufficient energy for growth. One water lily plant can typically reach 1 meter height (from the bottom) and cover a surface space with a diameter between 0.5 to 1 meter, while its flower can have a diameter of 3 to 6 cm. Water lilies are well adapted to their habitat. They grow and live on the edge of ponds and lakes, and in the shallow water parts. The climate of their habitat is usually warm and they live in water that is rich in oxygen and receives a lot of sunlight.

Plant growth in a pond depends on the ecological interactions and also is constrained by abiotic conditions (Brönmark and Hansson, 2005). Light conditions, water temperature, water velocity, and nutrient availability (e.g. phosphorus, nitrogen) and dissolved inorganic carbon concentrations are important environmental factors most often modelled to influence plant photosynthesis, respiration, washout and decay (Carr et al., 1997). However, in many cases for small lakes or ponds, conditions of light, water temperature and nutrients have a much higher contribution to the temporal dynamics than to the spatial dynamics of plant growth. The evolution and ecological interactions especially, are very important to the spatial pattern dynamics in a specific lake or pond (Brönmark and Hansson, 2005). Besides the influences from the initial seeding locations, spatial heterogeneity for rooted plants is due to factors like: (1) morphological factors, based on the size and growth pattern of the plants (e.g. cloned growth of the individual species); (2) environmental factors that are themselves spatially heterogeneous (e.g. bathymetry of the pond or lake); and (3) factors that permit the spatial arrangement of one species to affect the occurrence of another species through their interaction (e.g. intra- and inter- specific competition) (Dale, 2004). The contributions of the above factors to the plant growth have strong local effects in space. Specifically, in the case of a small pond, nutrients are often not limited, bathymetry has no big variation, and water is relatively still. Therefore, the influencing factors for water lily growth dynamics in this case mainly include: water temperature, light,

initial seeding positions, local interactions, biological aspects of plant, intra-
or inter- specific competition for space and light, etc.

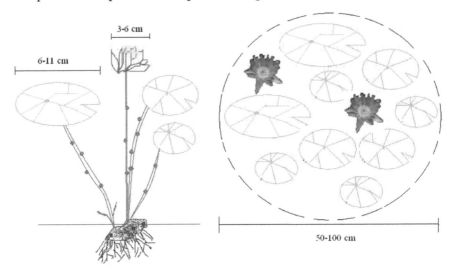

Figure 6-3 Spatial properties of water lily

### 6.3.3  Model setup

*Data preparation*

In order to setup and calibrate the CA model, an analysis of the conceptual
model structure was carried out for water lily growth dynamics. The main
data used were high resolution photos taken weekly by the department of
inland water systems at Deltares (WL|Delft Hydraulics). The photos show
the growth of macrophytes (water lily) at the water surface in a small pond
(about 52m×26m). The weekly photos taken during the year 2005 were used
for model setup and calibration. From the photos, it can be seen that water
lilies start to emerge at the water surface around week 16 in year 2005. Four
distinct periods of growth can be observed from the photos: (i) initial
growth, (ii) massive growth, (iii) saturated growth, and (iv) decay due to
winter weather conditions. Beside these weekly photos, the inputs to the
model also include the air temperature (obtained from airport
meteorological station at 10 km distance), duration of sunlight, and plant
properties, e.g. growth rate and life span.

The photos were preliminarily processed using Coral Paintshop ProX to
enhance the colours of interests (water lilies and water), as well as to reduce
the resolution to get smaller files. Then, the pre-processing results from
Coral Paintshop were imported into ArcGIS software for assigning the
values to water lily cells and water cells respectively. The resulting raster
was then converted to ASCII file format.

### Basic model configuration

Model development was carried out in the Matlab environment. The photo of week 18 was processed and taken as the initial matrix (Figure 6-4) for the basic model and for all other scenarios, which will be addressed later in this chapter. The general assumptions in this basic model are: (1) there are only two states (water lily with value 1 and water with value 2); (2) there is no nutrient limitation; (3) there is no influence from hydrodynamics and wind; and (4) spatial extension by seeding or by new leaves only happens in the nearest neighbouring cells, e.g. in case of the von Neumann neighbourhood scheme, only in the nearest 4 cells. (5) The neighbouring cells have equal rights to contribute to the growth of the central cell and rules are applied homogeneously throughout the whole domain.

The considered factors influencing the water lily spatial pattern dynamics for the basic model are: (i) weekly averaged water temperature and temperature differences between current week and previous week, (ii) weekly accumulated sunshine duration, (iii) plant properties (e.g. life span, age (weeks)), (iv) conditions in neighbouring cells. The considered neighbourhood schemes are Von Neumann and Moore neighbourhood schemes resp. Model cell size is based on the resolution of the photos. In this case, a cell size of 4 cm × 4 cm serves as the basic model grid. The model time step is considered to be one week.

Figure 6-4 Model initial condition retrieved from the photo of week 18

The rules are if-then rules based on the above influencing factors in combination with the available data and information: photos, water temperature, as well as properties of macrophytes obtained from expert knowledge. Some examples of simple rules are listed here: (1) *Growth rate*, the normal growth and the fastest growth are defined. For example, one of the rules for the fastest growth is that when the temperature is higher than

16 degrees and the sunshine duration is more than 40 hours a week, under such conditions, as long as there is at least one neighbouring cell that has a plant, a plant will grow in this cell as well; (2) *Life span*: a threshold of 18 weeks was considered: after a plant exists for 18 weeks, it will die; (3) *Spatial extension* of the plants: if there are favourable weather conditions, and at an age less than 18 weeks, and neighbouring cells are relatively empty - then the plant expands to its neighbouring cells either with the form of newly growing plants or by expansion of the leaves for existing plants, which is not specified further in this study; (4) *Mortality rate* varies with temperature, sunlight duration and temperature differences. For instance, if the temperature is between 15 and 12 degrees, and the temperature drop is more than 3 degrees compared to the previous week - then those cells which have less than 2 neighbouring cells with plants, will die. There are still many other empirical rules that are not listed here. Flow chart of modelling process can be found in Appendix II.

### *Scenarios for the effects of model configurations*

The selection of proper scales depends on many factors including the measurement data, the characteristics of the modelled system, the model structure and the user requirements. In this case, the cell size for the basic model was chosen to be 4 cm × 4 cm, which forms the basic scenario (S1) for this study. In order to select the proper cell size, scenarios were conducted with a multiplier of the original cell size from the basic model S1: S2 with 2 times (8 cm × 8 cm), S3 with 3 times (12 cm × 12 cm), S4 with 5 times (20 cm × 20 cm) and S5 with 10 times (40 cm × 40 cm) the original cell size (Table 6-1). In addition, modelling results are also influenced by the selection of the particular neighbourhood scheme. Therefore, these 5 scenarios are implemented both for Von Neumann and for Moore neighbourhood schemes.

Table 6-1 Scenarios for model cell configuration

| Scenarios | S1 | S2 | S3 | S4 | S5 |
|---|---|---|---|---|---|
| Cell size (cm$^2$) | 4×4 | 8×8 | 12× 12 | 20×20 | 40×40 |

### 6.3.4  Analysis of results

### *Basic model results*

The basic model, which has 4 cm × 4 cm grid size, is considered as the reference to check whether it is possible to depict the patterns exhibited on the photos by using a Cellular Automata model with simple if-then rules. The spatial pattern of week 30 is shown in Figure 6-5(a) for the Von Neumann neighbourhood scheme and in Figure 6-5(b) for the Moore neighbourhood scheme. The modelling results are compared with the photo in Figure 6-5(c) taken at that same week.

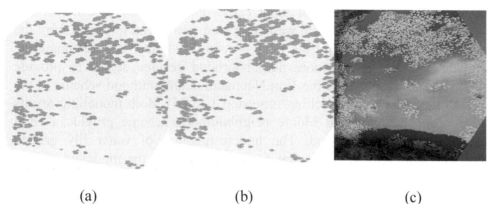

(a)                          (b)                          (c)

Figure 6-5 Spatial pattern predicted by basic model (a) with Von Neumann, (b) with Moore neighbourhood schemes vs. (c) photo for week 30

Since the pond is quite static and no nutrient limitation seems to be present, the spatial pattern of the modelling results evolving from this basic model are seen to agree quite well both in patchiness and in terms of total plant coverage, compared to the photo. However, some of the areas seem to have a lesser density compared with the photo. The resulting water lily coverage in the whole pond, together with the occupation percentages retrieved from the photos, is shown in Figure 6-6.

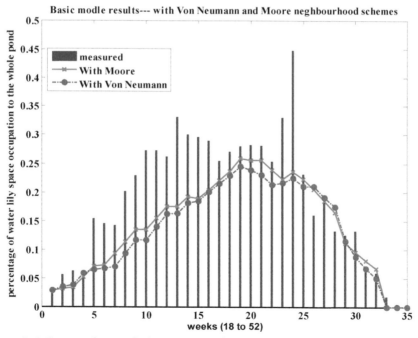

Figure 6-6 Comparison of plant occupation percentages: model results using Moore and Von Neumann neighbourhood schemes vs. retrieved from photos

Compared to the ratios of plant cells and total cells calculated from the weekly photos (percentage of plant cells), it is seen that the model can capture the growth trend quite well compared to the data retrieved from the photos, although in general the models seem to slightly underestimate the growth. The use of a Moore neighbourhood scheme seems to provide slightly better results than the Von Neumann neighbourhood scheme in the growth process (from modelling time step 1 to 25). Both modelling results using Von Neumann and Moore neighbourhood scheme provide a good match in the decay period. The underestimation of water lily growth indicates that there are also other factors influencing the modelled growth patterns. Therefore, sensitivity analyses for the models were developed to explore the effect of cell size of Cellular Automata in aquatic plant dynamics modelling.

### *Effects of model configurations*
The 5 scenarios listed in Table 6-1 were explored for both Von Neumann and Moore neighbourhood schemes. Figure 6-7 and Figure 6-8 show the comparison between photos and modelled water lily spatial patterns for different cell sizes for week 30. Figure 6-9 and Figure 6-10 show the comparisons of the percentages of water lily spatial occupations from model results with the data retrieved from photos. In addition, the resulting Root Mean Squared Errors (RMSEs) of percentages of water lily spatial occupation compared with measurements are shown in Table 6-2.

The modelling results indicate that not only the factors included in the basic model can influence the model results, but the choice of cell size and configuration can also be vital for the simulation of water lily growth dynamics. From the snapshots in Figure 6-7 and Figure 6-8, we can see that S1, S2 and S3 show quite similar spatial heterogeneity and the patterns obtained from S4 and S5 over-estimate the observations. In general, all scenarios captured the population increasing and decreasing trends, but the larger increase of the cell sizes over-estimated the plant coverage in the pond (Figure 6-9 and Figure 6-10). At the beginning and at the end of the calculations, the difference is rather small, but in the middle period of calculation when the plants exhibit massive growth, the resulting differences using large cell sizes are quite big. For the Moore neighbourhood scheme, the differences between different scenarios decrease when the cell size increases and the resulting percentages for S4 and S5 are closer to the measurements than the ones with the Von Neumann neighbourhood scheme. Besides, we see a decreasing trend in the data retrieved from photos between weeks 30 and 39 and a dramatic increase in week 41, which may not be very accurate based on expert knowledge obtained for real situations. It indicates that the data retrieved from photos need to be verified in the further study.

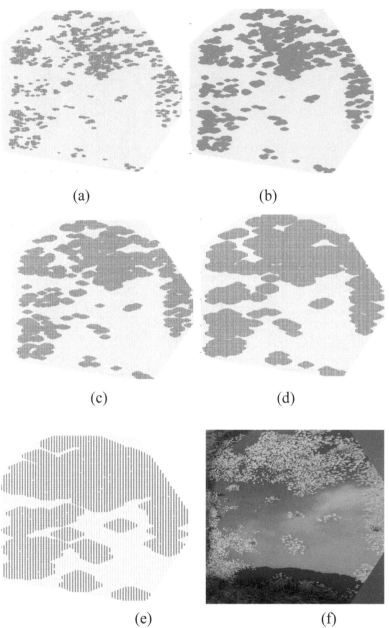

Figure 6-7 Comparison of the snapshots (week 30, Von Neumann) of model results: scenarios:(a) S1, (b) S2, (c) S3, (d) S4, (e) S5 with the photo of week 30 (f).

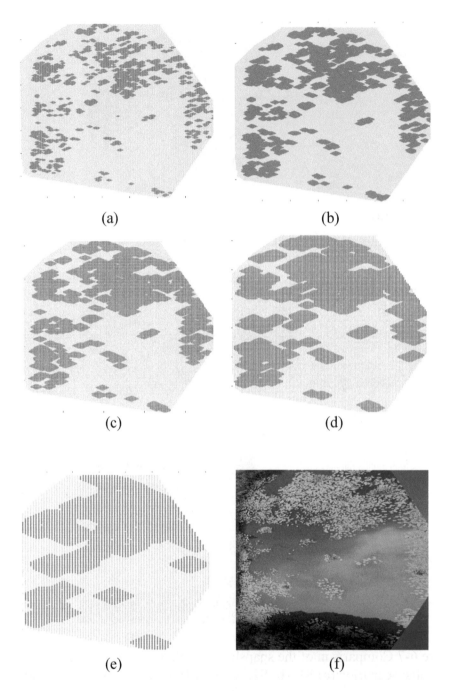

Figure 6-8 Comparison of the snapshots (week 30, Moore) of model results: scenarios  (a) S1, (b) S2, (c) S3, (d) S4, (e) S5 with (f) the photo of week 30.

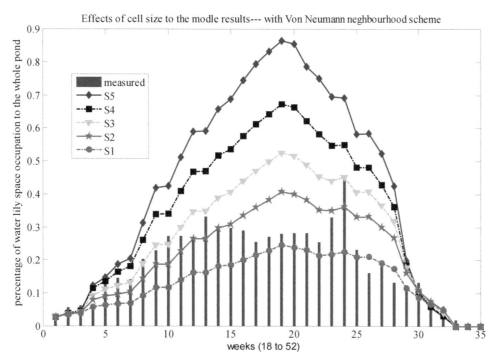

Figure 6-9 Modelled Water Lily spatial occupation with different cell sizes using Von Neumann neighbourhood scheme

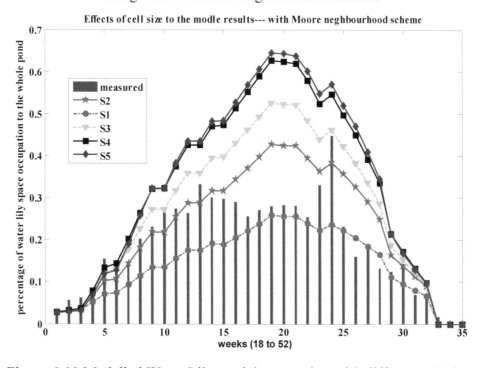

Figure 6-10 Modelled Water Lily spatial occupation with different cell sizes using Moore neighbourhood scheme

Table 6-2 RMSEs of resulting water lily spatial occupations compared to the photo of week 30

| Scenarios | S1 | S2 | S3 | S4 | S5 |
|---|---|---|---|---|---|
| RMSE with Von Neumann | 0.082929 | **0.07276** | 0.122442 | 0.201828 | 0.306644 |
| RMSE with Moore | **0.074576** | 0.076461 | 0.123024 | 0.178997 | 0.189682 |

Still, if we consider the root mean square error as another measure for the differences between modelled plant occupation and the data retrieved from photos (Table 6-2), we also see differences from different scenarios and with different neighbourhood schemes. As an example, we take the resulting RMSEs of the model results for week 30 with both Von Neumann neighbourhood scheme and Moore neighbourhood scheme. Based on RMSEs, the best scenario is S2 with Von Neumann neighbourhood scheme which has the lowest RMSE among all scenarios.

Both from the analyses of cell sizes and neighbourhood schemes, based on the comparisons of the spatial patterns and resulting errors, we see that different neighbourhood schemes and different cell sizes can obviously influence the modelling results. S1 and S2 seem to give better modelling results compared to other scenarios. This could be explained as follows. A mature water lily leaf can have diameter of about 6 to 11 cm (Figure 6-3). If we choose the cell size of 4 cm (S1) or 8 cm (S2), it covers a range of a leaf size. In terms of different neighbourhood schemes, if a similar region is covered, Moore requires a smaller cell size than Von Neumann because it covers more cells (9) than Von Neumann (5).

This leads to the general conclusion that in individual-oriented plant growth modelling, the model performs best when the selected neighbourhood configuration has the characteristic dimensions of the particular plant species.

### 6.3.5 Conclusions and discussion

The research in this chapter aimed to explore the capabilities of the Cellular Automata concept for detailed modelling of water lily plant growth, using high resolution photography in a confined pond as a reference for detailed comparison. The case study showed that a fine-scale Cellular Automata model in the pond environment is quite capable to capture the individual behaviour of water lily growth including seasonal variations and patchiness characteristics. It also shows that high resolution photography is a cheap and practical source of data for verifying plant population dynamics on the water surface (provided a stable platform for taking photos is available) to obtain as was the case for the local pond in this research. Detailed patchiness and local patterns in the smaller scales can be obtained when using a CA model where the selected cell sizes are of the same magnitude as the represented plant biological characteristic dimensions. In general, it is vital to develop proper geometrical CA rules to model any real biological system.

Clearly, retrieving information from photos is subject to the modeller's expertise as well as to the quality of the photos. Therefore, not only model configuration, biological information and expert knowledge, but also accuracy of measurement data (photos in this case) is vital. Underwater photography could perhaps be used for subsurface plant growth. The extension of this study from relatively small stagnant ponds to lakes or bigger water bodies needs to include the influence from hydrodynamics (e.g. wind-driven currents and waves).

## 6.4 Summary

This chapter focused on the use of Cellular Automata (CA) in revealing spatial pattern dynamics in aquatic plant modelling. For different types of CA models, it was shown how the transition rules are derived, which can be if-then rules, stochastic rules, automatic learning rules as well as fuzzy inference rules, respectively. One case study was carried out using a traditional CA approach based on empirical knowledge and measurement data from high resolution photography. This case study showed the applicability of CA based modelling in revealing individual-oriented fine scale spatial pattern dynamics. It also showed that when CA based models capture the characteristic length of an individual plant, very fine spatial patterns observed in high resolution photos can be well captured by a CA-based model. CA-based models not only can be used for simulating individual-oriented fine scale spatial pattern dynamics, they also prove applicable for larger scale applications (Mynett and Chen, 2004).

One of the advantages of using CA is that it is not sensitive to instability problems so that it can model phenomena that occur on fine spatial scales as well as long time scales, e.g. the growth dynamics of macrophytes considered in this research. However, in general, one does need to consider the resident times and flow influence when more processes with different scales take place. This specific test case showed that CA can be used at very fine scale simulations of individual plants. In this sense, the present study of individual water plants can be seen as a detailed experiment, showing the validity of CA at very fine scales. However, in general, there is hardly enough fine scale spatial data available and the interest may be on large-scale biomass accumulation of substrates or densities of collective individuals, in which case larger computational cells may be sufficient.

# Chapter 7

# Revealing spatial pattern dynamics in aquatic ecosystem modelling with multi-agent systems[4]

## 7.1 Introduction

In the field of Environmental Hydroinformatics (Mynett, 2002; Mynett, 2004; Mynett and Morales, 2006), a range of modelling paradigms can be distinguished: physically-based modelling (e.g. for numerical flow simulation ), discrete modelling (e.g. Cellular Automata (Mynett and Chen, 2004; Li et al., 2006a), and agent based approaches (e.g. individual / agent based model (Morales et al., 2006) or recently developed: Multi-Agent Systems (Ferber, 1999; Li et al., 2009c)). The latter one will be explored here to simulate and forecast the growth, decay and spreading of macrophytes.

Numerical simulations are well established in hydrodynamics and water quality modelling often based on partial differential equations solved by particular numerical schemes (Stelling, 1984; Postma, 2007; Los et al., 2008). Cellular Automation (CA) is a discrete modelling paradigm based on local interactions on a regular grid of cells, as introduced before. It can deal with spatial interactions between populations and their local environment. CA is commonly applied to a discrete spatial domain in a fixed lattice structure where the future state of each cell is determined by the cell state itself plus the states of immediate neighbourhood cells through transition rules.

Individual-Based Models (IBM) and Multi-Agent Systems (MAS) can represent an individual entity or a particular class. This seems to be relevant for macrophytes whose growth, decay and spreading is collective. MAS is a collection of interacting agents that can couple different components within a model considering both local interactions and individual behaviour in a very flexible way that seems more applicable in terms of a better representation of reality and practical use (Ferber, 1999). Therefore, a multi-

---

[4] Based on:
Li, H., Mynett, A.E., Qi, H. and Penning, E., 2009, Multi-Agent Systems in modelling aquatic population dynamics in Lake Veluwe, Netherlands. submitted to Journal of Ecological informatics.

agent system approach is used here for simulating macrophytes' collective behaviour as a discrete representation in the spatial domain.

This chapter focuses on multi-agent systems and the use of multi-agent systems in the field of spatial pattern dynamics modelling for aquatic macrophytes' growth. Section 7.2 introduces the concept and development of MAS in general, which includes the background, the components involved in MAS as well as some past experiences with using MAS in aquatic ecosystem modelling. Section 7.3 describes a case study developed in this research for the modelling of population dynamics in Lake Veluwe, the Netherlands. It specifies in more detail the conceptual model, the model implementation as well as the analyses of results based on the GIS maps of population density. Section 7.4 summarizes the applicability of MAS and what MAS can offer in the spatial dynamics modelling of aquatic populations.

## 7.2    Multi-agent systems

### 7.2.1   Concept

First of all, what is an agent? In the physical / biological context, an agent can be an individual insect, a plant, a fish, or a bird; in the social context, it can also be a farmer or an organization, or a group of people. Ferber (Ferber, 1999) gave a definition of an agent and of a multi-agent system which is widely accepted by many branches of research:

> *An agent can be a physical or virtual entity that can act, perceive its environment (in a partial way) and communicate with others is autonomous and has skills to achieve its goals and tendencies. It is in a multi-agent system (MAS) that contains an environments, objects and agents (the agents being the only ones to act), relations between all the entities, a set of operations that can be performed by the entities and the changes of the universe in time and due to these actions.*

Such "agents" have some common features (Gilbert and Troitzsch, 2005):
- *autonomy*: agents operate without others having direct control of their actions and internal state
- *social ability*: agents interact with other agents through certain rules
- *reactivity*: agents also interact and respond to the environment;
and some of the agents also have:
- *pro-activity*: agents are able to take initiative, engaging in goal-directed behaviour.

A computational agent-based model is based on the above agents and their interactions in order to simulate system patterns at a higher scale and, most likely, with emergent behaviour.

A multi-agent system is a system composed of many intelligent agents, which can be seen as a group of agent-based models. MAS originated from distributed Artificial Intelligence (DAI), a sub-field of Artificial Intelligence (AI) (Gilbert and Troitzsch, 2005). MAS solves problems which are difficult or impossible for an individual agent to solve.

Besides the considerable developments in software toolkits for ABM and MAS, there are also other important reasons for the rapid development of these types of modelling: (a) the models have emergent properties: the whole is more than the sum of its parts because of the interactions between these parts; (b) they have a better way of describing natural systems. For example: it is more natural to describe how *actual* fish move in a lake than to use equations for the *average* density of fish.

## 7.2.2  Components in MAS modelling

In a multi-agent system, the agents are presumed to be acting in what they perceive as their own interests like reproduction, social status, competition, adaptation and even learning. Multiple agents operate simultaneously in order to reproduce and/or predict the actions of complex phenomena.

There are mainly two actors in MAS modelling: (i) agents and (ii) their environment. As shown in Figure 7-1, for the implementation of MAS in a computer program, the environment is often discretized into 2-dimensional lattice consisting of many "substance" layers to represent the influencing factors to the agents (e.g. nutrients and hydrodynamics as environmental factors for aquatic plant growth) and one agent position layer. Each agent has its own processes, e.g., growth, death, interaction, spatial extension and energy gain or loss by motion or feeding under the given environment. There are interactions between agents especially when they are close to each other in space, e.g. they may need to compete for food or other types of energy, or one type of agent is the predator or grazer of another type of agent. There are also interactions between agents and environmental factors, which can be the constraints from environment to agents, and there are also feedbacks from the behaviour of agents to their environment.

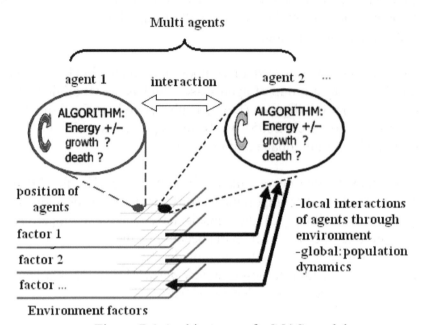

Figure 7-1 Architecture of a MAS model

### 7.2.3  Advantages and possible problems in MAS modelling

Based on Sycara (Sycara, 1998), Ferber (Ferber, 1999), Lesser (Lesser, 1999), and Vlassis (Vlassis, 2007), the advantages of using MAS are summarized to be:

- making observations and performing actions at multiple locations simultaneously, and therefore taking advantage of geographical distributions.
- providing a framework that allows for interactions at different scales and the simulation of emergent ecosystem properties.
- providing a natural way of representing the underlying mechanism of the real world system.
- efficiently retrieving, filtering, and globally coordinating information from rules that are spatially distributed.
- enhancing overall system performance, specifically along the dimensions of computational efficiency, reliability, extensibility, robustness, maintainability, responsiveness, flexibility, and reuse.
- able to consider both quantitative and qualitative parameters, and having the capacity to integrate quantitative variables, differential equations and rule based behaviour into the same model.
- computer programming, which can be done in any language, but Object Oriented Programming (OOP) being the most appropriate

method since the concept of an object is similar to the concept of an agent.

Still, we need to be cautious in using ABM and MAS because they are computationally intensive and time consuming when the model reaches a certain level of detail. Such problems often exist but are becoming less important because of the continuous development of computational power.

## 7.2.4  MAS model construction

It is very important to design MAS models systematically that show both agents' autonomous and collaborative properties. Based on (Park and Sugumaran, 2005), several steps have to be followed to develop a MAS model (Figure 7-2): problem analysis, conceptual model development and MAS model implementation. Some considerations in the problem analysis phase include: (a) suitability assessment of MAS for the specific problem; (b) user requirement specification; (c) main goal and sub-goals definition of the MAS model based on user requirements; (d) availability of data and information. Before the implementation of MAS with the selected software toolkit, developing a conceptual model is vital to identify the detailed requirements, which is the most important and also the most difficult aspect in constructing a MAS model. First consideration in the conceptual model is the mapping of the goals to agents (Park and Sugumaran, 2005), following up by detailed description of agent internal architecture (agent autonomy) and system external architecture (agent collaborative behaviour). If a proper conceptual model is designed, the MAS model can be easily implemented in a selected software toolkit with considerations of the model reliability and adaptability, etc. In summary, ABM or MAS type models can be of use (Bonabeau, 2002) when:

1. interactions between agents are complex, non-linear, discontinuous, or discrete;
2. space is crucial and the agents' positions are not fixed;
3. the population is heterogeneous, when each individual is (potentially) different;
4. the topology of the interactions is heterogeneous and complex;
5. the agents exhibit complex behaviour, including learning and adaptation.

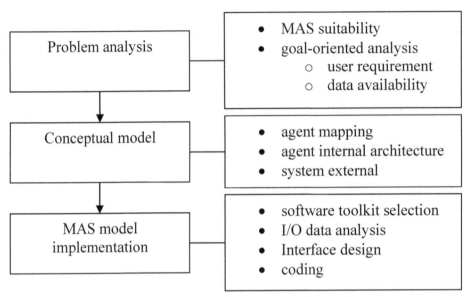

Figure 7-2 Architecture development process for multi-agent systems
(revised based on (Park and Sugumaran, 2005))

In the field of aquatic ecosystem modelling, the data is often very limited
and it is of importance to use all possible data sources in constructing MAS
ecosystem models (Figure 7-3). The main information needed for aquatic
ecosystems is the domain knowledge from biologists and ecologists. This
part of information is the core for setting up the MAS conceptual model. In
order to represent the spatial pattern dynamics, spatial information needs to
be included explicitly. GIS maps and Remote Sensing Imageries, as well as
high-resolution photos become more and more popular in many fields. The
drawback of using such kind of information, especially remote sensing data,
is that it hardly supplies complete time series of information over a long
period of time, but this can be compensated by adding long term point
measurements. Furthermore, some factors which have less spatial
heterogeneity can be sufficiently represented by point measurements.
Finally, the most direct way of combining many different types of models is
to combine the outputs of different component models as the input data for
the MAS model.

The main outputs of the MAS model are the spatially explicit global
population dynamics, which emerge from the multi-agent behaviours.

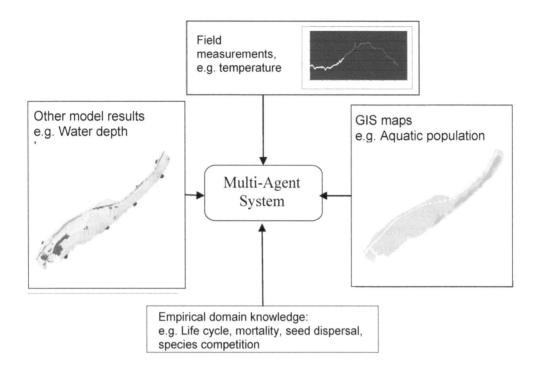

Figure 7-3 Data and information sources for constructing MAS model

One of the most important reasons for the rapid expansion of ABM and MAS as modelling paradigms is that software toolkits released within the last few years have made agent-based modelling easy enough to be attractive to many researchers in a variety of subject areas. Among these toolkits, the most significant ones are Swarm, Repast, Netlogo, Java Swarm, and MASON, etc. (Railsback et al., 2006) summarized and compared the above most popular software toolkits for ABM simulation, and concluded that Netlogo was the preferable toolkit for their ABM and MAS model. In terms of accessing external spatial data from GIS or ASCII files, and linking with other modelling tools (e.g. Matlab, R and Weka), the Repast Simphony (http://repast.sourceforge.net/) can be used. In many cases, Matlab was also used as alternative tool to develop agent based model (Morales et al., 2006). The main reasons of using Matlab are: it has enormous mathematical functions available to use and it can deal with relatively large matrices easily. Moreover, versions after 2008 also support object-oriented programming.

## 7.2.5  MAS in aquatic ecological modelling

MAS has developed rapidly during the past decades and is being applied in many different fields. Most of the applications are in the fields of social systems and organizations. (Davidsson et al., 2007) summarized the past applications of MAS and only 2 out of 33 of the literatures were related to ecological systems: one on Plio/Pleistocene Hominid food sharing and another one was on plant disease incursion management. There are some applications of MAS in aquatic ecological modelling (Weber et al., 2006). Still, the characteristics of aquatic ecosystems make it a potential application area. The reasons of only few applications may be due to the lack of knowledge and detailed information as well as the difficulties in model validation. In aquatic ecosystems, the main agents can be algae, plants, fish, ducks etc, and they live in water, which means they are constrained and even dominated by the water body. Therefore, the contents in the water body and the motions of the water body have high influence on the living creatures in the aquatic ecosystems. The construction of MAS models for aquatic ecosystems therefore involves many factors of both the characteristics of the water body (water depth and velocity, nutrient concentrations, light penetration, oxygen etc.), and the characteristics of the food-web in the specific area.

## 7.3    Multi-agent system for Lake Veluwe, Netherlands

### 7.3.1  Introduction

***Case study area***
Lake Veluwe (Veluwemeer) is an artificial lake in the Netherlands (Figure 7-4) stretching from South-West to North-East with a surface area of over 30 square kilometres and an average depth of 1.55 m. It is currently a macrophyte-dominated system. Since the late 1960's and early 1970's, submerged vegetation was affected by eutrophication, which led to the first big shift of vegetation in this lake from varieties of macrophytes to only sparse patches of submerged *Potamogeton pectinatus* (*Pp*). In 1979, measures were taken to reduce excessive phosphorus loading (Hosper, 1997; Van den Berg et al., 2001), which gradually increased the quality of this lake. Between 1987 and 1993, the dominance of *Potamogeton pectinatus* (*Pp*) decreased, while Charophyte meadows (e.g. *Chara aspera* (*Cs*)) expanded during the same time interval. Nowadays, seasonally persistent clear patches associated with the Chara meadows are observed. The pattern changes of the dominant macrophytes might have resulted from the change in underwater light climate (Coops and Doef, 1996; Van den Berg et al., 1998).

*Pp* and *Cs* can be used as indices for the quality of a water body. They can coexist in a water body with good water quality, and their growths are influenced by light, nutrient and optimal growth positions, etc.

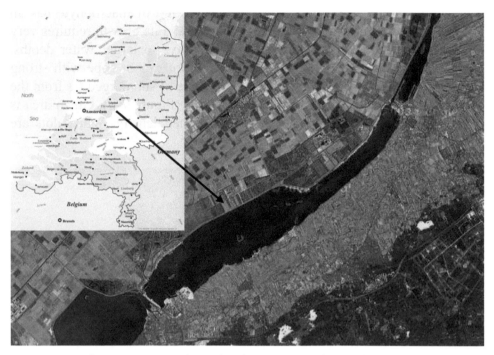

Figure 7-4 Location of Lake Veluwe, the Netherlands

## Background biological knowledge

***Dominated plants*** Currently, *Chara aspera* (*Cs*) and *Potamogeton pectinatus* (*Pp*) are the main macrophytes in Lake Veluwe. *Potamogeton pectinatus* (*Pp*, *Schedefonteinkruid* in Dutch) is a submerged annual growth macrophyte. It can grow up to two meters, depending on water depth and quality. It often forms most of its biomass in the top layer of the water column to optimize light capturing. Its rhizomes are tuber-shaped, so it mainly absorbs nutrient from the soil. The extension of tubers is not very obviously affected by the flow factors, but by animals, other plants and human actions. *Pp* can invade in the aquatic environment since it can deal with relatively stronger waves and currents and live in less clear water, compared to *Cs* (Kantrud, 1990). The population reproduction of *Pp* relies on the tubers formed in the previous year, so that the amount of previous year's vegetation leads to 70% of the new growth. Some 20% of new growth comes from nearby formed new tubers (adjacent cells). Seed dispersal by *Pp* is quite slow (less than 6% recorded by Li, et al. 2002 (Li et al., 2002)) compared to *Cs*.

*Chara aspera* (*Cs*) is a rooted highly developed macrophyte. The plants grow mainly in alkaline freshwater lakes and ponds. They propagate by formation of very small (0.5-0.8mm) seeds: diploid oospores. Oospores spread via flow and water birds since at shallow areas about 50% of the biomass was consumed enhancing the potential dispersal of oospores by water birds (Van den Berg et al., 2001). This kind of macrophyte has an average height of about 22 cm in Lake Veluwe, and its growth requires very good light conditions and, therefore it is sensitive to larger water depths. The roots of *Cs* are not very strong and therefore cannot cope with strong flows and waves. We can see the spatial coverage of submerged *Cs* from the photo taken in the airplane above Lake Veluwe (Figure 7-5). Green area is covered by *Cs* under water, darker means higher density. White dots are swans preferring this lake as their habitat.

Figure 7-5 Submerged *Cs* observed in the photo taken in an airplane above Lake Veluwe (green area in the lake is covered with submerged *Cs*)

The properties of these two different plants are summarized in Table 7-1.

***Interactions between macrophytes Pp and Cs*** can occur due to the availability of nutrients, light, space as well as their biological properties, which are assumed here as intraspecific and interspecific competition. The intraspecific competition is the competition for food and space within the same species, while the interspecific competition is between different species. Based on (Van den Berg, 1999), *Pp* is a better competitor for light and temperature than *Cs* due to the shading effects and capability of nutrient storage with an earlier growth under lower temperature. Whereas, *Cs* has much shorter life span than *Pp*, and *Cs* spreads seeds (can be 1 million per m$^2$) and germinates much more than *Pp*. Besides, based on the experiments in (Van den Berg, 1999), the growth of *Cs* was associated with Dissolved Inorganic Carbon (DIC) depletion. At lower DIC (high pH), *Cs* has higher photosynthetic rate than *Pp* therefore leading to an even higher *Cs* concentration. Strong decrease of DIC was observed in Lake Veluwe especially in spring and summer period (Van den Berg, 1999). Therefore, in the area dominated by *Cs*, *Pp* can hardly grow.

Table 7-1 Properties of *Cs* and *Pp*

| | *Pp* | *Cs* |
|---|---|---|
| **Biology** | | |
| length | max 2m | max 22cm |
| life span | yearly | about 3 months |
| seeds | tuber mainly (rooted) | oospores (mainly, and large amount) and bulbils (small rooted seed, small part) |
| root | strong, big | weak and small |
| predation | by ducks, birds and fish | by ducks, birds and fish |
| **Environment** | | |
| water depth | can be deeper till 4m | cannot be deeper than 2m |
| light | can withstand lower light | high light attenuation |
| flow | can live in relatively strong wave area, leaves flow to flow direction | strong flow takes away their roots, hard to live. |
| pH | less grow with high pH | more grow with high pH |
| water quality | good water quality, absorb nutrients from soil | good water quality |

*Environmental factors*

Many environmental aspects affect the growth of *Pp* and *Cs*, which in turn affect the environment. The main environmental factors include the considerations of flow conditions, nutrient availability, light penetration, and water depth, etc. for the plant growth in this lake. There is a direct relation between seed dispersal and flow direction and velocity. If the flow is from left to right, the biggest probability of seed spreading is to the right. In addition, the distance that the seeds can spread is linked to their fall velocity and the flow velocity. The smaller the flow velocity, the higher the fall velocity, the closer the seeds remain to their source plants. With larger flow velocities, the seeds spread further and are dispersed more.

Nutrients are the food for plant growth and they can directly constrain the growth speed and growth quality of plants. A suitable nutrient concentration can offer a good growth environment. However, in water bodies with large plant coverage (which absorb the nutrients and then nutrients become limited in the water body for the phytoplankton to grow) eutrophication does not become an issue.

The photosynthesis of any kind of plant needs light, especially for submerged water plants. The amount of light that reaches the sediment surface and the submerged plants depend on the transparency of the water, which in turn depends on the amount of suspended organic matter (e.g. phytoplankton) and sediment particles in the water column, as well as the colour of the water itself.

### *Measurements*
The possible data sources include point measurements and GIS maps, Remote Sensing images etc. In this research, GIS data and point measurements are the main sources of measurement data. GIS data supply both the model domain of the lake Veluwe and the yearly densities of macrophytes. Figure 7-6 shows the shape of Lake Veluwe and the measured plant densities of *Cs* and *Pp* in different years (1994, 1997 and 1999) in Lake Veluwe. The darker areas are covered by plants and different colour depths imply different plant density.

Point measurement data represented as daily water temperature and extinction coefficients, were downloaded from http://www.waterbase.nl/. In general, extinction coefficients show a decreasing tendency since 1993. Nowadays, the water quality in this lake is becoming better and more and more suitable for *Cs* to grow. Water depth data were supplied by the Dutch National Water board (Figure 7-7).

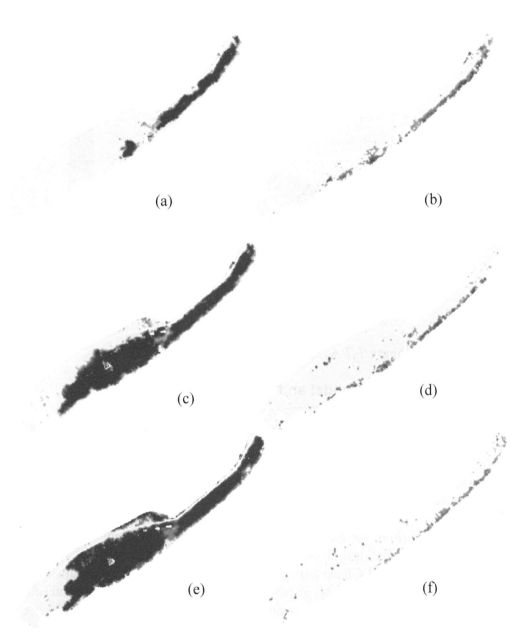

Figure 7-6 *Cs* and *Pp* densities and distributions in year 1994, 1997 and 1999 (a) *Cs*: 1994, (b) *Pp*: 1994, (c) *Cs*: 1997, (d) *Pp*: 1997, (e) *Cs*: 1999, (f) *Pp*: 1999

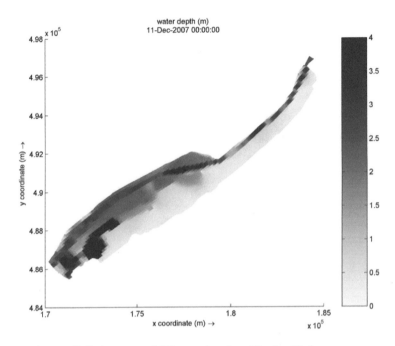

Figure 7-7 Averaged Water depth of Lake Veluwe

## 7.3.2  Conceptual model and operational rules

In order to construct a proper model for Lake Veluwe, a MAS model needs to include many aspects, e.g. agents and their biological knowledge, environmental factors, and the interactions between agents as well as between agents and their environment. The aspects in the MAS model design are:

*Simulated entities or multi-agents*: there are two different macrophytes considered in the design of this MAS model, which are *Potamogeton pectinatus* (*Pp*) and *Chara aspera* (*Cs*). Every agent has its own attributes, e.g. life spans, individual growth function, spatial spreading and seed spreading. Water birds are only considered as influencing factors in the seed dispersal process of *Cs* in this case.

*Spatial Position*: refers to the assumption of a particular location in the physical space for the simulated entities. This can be expressed as either absolute distance or relative positions between entities.

*Communication*: the entities can have some or no interactions with one another. The interactions take place in the form of intra-agent communication and inter-agent communication.

*Mobility*: refers to the ability of an entity to change position in physical space. In this research, the mobility of plants is in the seed dispersal. Mature plants disperse seeds driven by the external environmental factors, in this case, mainly flow.

*Environment*: agents live in such an environment that is designed as the background for the MAS model. This environment interacts with the agents and all the agents will join in the campaign. In association with the system elements in the MAS model, the processes and factors involved in the growth of submerged plants *Pp* and *Cs* include germination, individual growth, spatial extension, seed dispersal and mortality. All of the above processes are influenced and constrained by the external environmental factors, including temperature, nutrients, water depth and water motion, etc.

### *Assumptions*

Some assumptions are summarized below based on both literature review and expert knowledge:

- The states for each agent are: density, life span, life ability, height, and age, which change dynamically in time and space;
- Important external factors are: extinction coefficient, water depth, flow velocity and water temperature. Users can define the thresholds or functions of the relationships between above environmental factors and macrophyte growth;
- Competition occurs between macrophytes (same or different species) co-existing in the same environment. They compete over resources, nutrients, space, lights, etc.);
- In each environmental background cell, more than one type of agents exist and they can compete and coexist based on the underlying conditions;
- The maximum life span (*MaxLifeSpan*) for *Pp* is 38 weeks and for *Cs* is 13 weeks;
- The maximum growth length is taken as 2 meters for *Pp* and 20 cm for *Cs*;
- The same type of plants in each computational cell have the same properties, same life stage, same height, same life span, which is lumped as one agent;
- Winter is skipped from the model since plants do not grow in that period, which is represented in the way that seeds do not have time delay to grow;
- Randomness is added into initial age distribution, density variations and seed dispersal;

- Seeds for both plants germinate only in the range of 4-25 $^{o}C$ (Li et al., 2002).
- Seed germination rate for *Cs* is taken as 15% (Van den Berg et al., 2001), and for *Pp* is 6% (Li et al., 2002).

## *Processes*

Future states $S_{t+1}$ (height, density (*D*)) of the plant depend on the previous states $S_t$ (e.g. height and density) and external constraints from light, temperature, water depth and water motion. The measurements and experimental analysis in Van den Berg (Van den Berg et al., 1999) showed that at certain measurement locations, macrophytes in Lake Veluwe emerge around the middle of April and reach their saturation density at the beginning of July. After the end of September, the vegetation cover starts to decrease due to both temperature drop and bird grazing.

Beside the individual growth of plants, there are always interactions and communications between plants and their environment, which lead to variations in aquatic plant density. The change of aquatic plant density within certain environmental computational cells is described briefly in Table 7-2. For two species growth, the Lotka-Volterra (LV) equations are widely used (Jørgensen and Bendoricchio, 2001), which considers a logistic growth for each species by adding a limiting term from another species due to resources competition. However, for the ones which compete/interact not only because of resources, the formulation for competition and interaction need to be species-specific. Extending from the general form of LV equations, density variation $\dfrac{\partial D}{\partial t}$ for certain species on certain location is determined not only by macrophytes' individual growth $\dfrac{\partial D_g}{\partial t}$, moreover, as shown in Table 7-2, it also depends on seed germination $\dfrac{\partial D_{sg}}{\partial t}$, spatial interaction and extension $\dfrac{\partial D_e}{\partial t}$, interspecific and intraspecific competition $\dfrac{\partial D_i}{\partial t}$, morality $\dfrac{\partial D_m}{\partial t}$ as well as seed dispersal $\dfrac{\partial D_{sd}}{\partial t}$. Furthermore, a small random variation (*R*) needs to be considered to represent the stochasticity in aquatic ecosystem. The items shown in Table 7-2 are described separately hereafter.

Besides, a parameter called life ability (*life_a*) is introduced. *Life_a* in this research means that every plant has an ability that refers to its lifespan. For example, assuming the ideal *life_a* is *s* and a plant can live 100 days in an optimal environment, but bad conditions can make this plant live for only 20 days, then the life-ability for this plant becomes s/5. We use *life_a* as a parameter to measure the lifespan of the plant that may suffer from changes and constraints from environments as well as disturbances from human beings.

Table 7-2 Density variation for the coexistence of *Pp* and *Cs*

| Density function | $\dfrac{\partial D}{\partial t} = \dfrac{\partial D_g}{\partial t} + \dfrac{\partial D_{sg}}{\partial t} + \dfrac{\partial D_e}{\partial t} - \dfrac{\partial D_i}{\partial t} - \dfrac{\partial D_m}{\partial t} + \dfrac{\partial D_{sd}}{\partial t} + R$ |
|---|---|
| $\dfrac{\partial D_g}{\partial t}$ | individual plant growth: $f(r,D,K,t)$, $K$ is carrying capacity, $r$ is growth rate. |
| $\dfrac{\partial D_{sg}}{\partial t}$ | seed germination: $f(D,R_g,t)$, $R_g$ is the probability of germination; change of seed germination depends on current density, and computational time of the year |
| $\dfrac{\partial D_e}{\partial t}$ | spatial extension: $$f(D_{(i,j)}, D_{(i-1,j)}, D_{(i,j-1)}, D_{(i+1,j)}, D_{(i,j+1)}, P...)$$ which is based on Cellular Automata concept, depending on neighbouring area density conditions and a random probability of extension $P$ |
| $\dfrac{\partial D_i}{\partial t}$ | interspecific and intraspecific competition: $f(D_1, D_2, age, height, t)$, where: *age* and *height* are the properties of two plant species. |
| $\dfrac{\partial D_m}{\partial t}$ | mortality: $f(D_1, D_2, temp, age, t)$, where *temp* is temperature, *age* is plant age that cannot be more than maximum age of the plant. |
| $\dfrac{\partial D_{sd}}{\partial t}$ | seed dispersal: $$f(D, v, age, age_{(i,j)}, age_{(i-1,j)}, age_{(i,j-1)}, age_{(i+1,j)}, age_{(i,j+1)}, t, P_1, P_2),$$ which depends on cell density, age, flow velocity (v) and neighbouring plant age, a small random factor $P_1$ as the probability of seeds from neighbouring regions dispersing to the cell, as well as $P_2$, a random probability of growth of dispersed seeds those are added into total density of certain location. |
| $R$ | a small random variation considered to represent the general stochasticity in aquatic ecosystem |

### Germination

The two plants considered here have quite different germination properties. *Pp* largely relies on previous year's plant (more than 70 percent) since the regeneration of *Pp* populations is mainly from the root tubers and rhizomes of last year's plants. *Pp* also has seeds, however, less than 6 percent of them can germinate in an optimal environment (Li et al., 2002), which does not contribute much to their population regeneration. *Cs* disperses large numbers of oospores seeds with about 15% germination rate (Van den Berg et al., 2001). It also has so called bulbils as seeds as part of germination adding to the oospores seeds.

*Growth*

Logistic growth equation (Verhulst, 1838) is a common model of population growth, where the rate of reproduction is proportional to: (a) the existing population, and (b) the amount of available resources. The second term models the competition for available resources, which tends to limit the population growth. The logistic model is formalized by the differential equation Eq. 7.1:

$$\frac{dD}{dt} = rD(1 - \frac{D}{K})$$
(7.1)

where *r* defines the growth rate and *K* is the carrying capacity, *D* represents population density size and *t* represents time. In ecology, the solution to the equation (with $P_0$ being the initial population) is:

$$D(t) = \frac{KD_0 e^{rt}}{K + D_0(e^{rt} - 1)}$$
(7.2)

where, $\lim_{t \to \infty} D(t) = K$, which is to say that *K* is the limiting value of *D*: the highest value that the population can reach given infinite time (or come close to reaching in finite time). It is important to stress that the carrying capacity is asymptotically reached independently of the initial value $D(0) > 0$, also in case that $D(0) > K$.

Therefore, if we assume previous population density *D(t-1)* as the initial value for the calculation of current population density, Eq. 7.2 becomes 7.3:

$$D(t) = \frac{KD_{t-1} e^{r}}{K + D_{t-1}(e^{r} - 1)}$$
(7.3)

In this case, nutrients are sufficient to let both *Pp* and *Cs* reach maximum density. Due to the importance of temperature for macrophytes' growth, growth rate *r* is defined based on seasons, temperature (*T*) and weekly temperature differences (*WTD*). The analysis of relationships between population and growth rate can be found in Appendix III. The chosen values for *r* (*r_cs*: growth rate for *Cs*; *r_pp*: growth rate for *Pp*) in the MAS model is shown in Table 7-3.

Table 7-3 Temperature dependent growth rate

| growth rate | Jan.-Aug. | | | Sep.-Dec. | | |
| --- | --- | --- | --- | --- | --- | --- |
| | $t \geq 10$ | $6 \leq t < 10$ | $t < 6$ | $t \geq 10$ | $t < 10$ | |
| | | | | | $-2 \leq WTD < 0$ | $WTD < -2$ |
| r_cs | 0.7 | 0 | 0 | -0.7 | -1.5 | -3 |
| r_pp | 0.5 | 0.5 | 0 | -0.5 | -1.2 | -2.5 |

Individual plant heights change with a linear function of plant age (Eq. 7.4).

$$Height(t) = \frac{maximum\_height \times plant\_age(t)}{maximum\_lifespan} \qquad (7.4)$$

where, *maximum_height* is the maximum plant height for *Pp* or *Cs*, *plant_age* is the current age of the plant, *maximum_lifespan* means the maximum life span each type of plant can reach.

*Species interactions*
The two types of plants considered here can influence each other by competition and coexistence. The main competitions in Lake Veluwe between *Pp* and *Cs* are space occupation, seed dispersal and germination, as well as growth rate. Nutrients are not considered in this case since nutrients in this lake are not limiting for both plants. *Pp* grows taller than *Cs* and when it grows close to the water surface, it swings with the flow, which reduces light intensity for *Cs* to grow. Therefore, the growth sequence determines the influence from *Pp* on *Cs*. The MAS model considers that only when *Pp* grows a few weeks earlier than *Cs* (e.g. 4 weeks), the growth of *Cs* can be limited by *Pp* with reduced life ability and reduced growth rate. In this lake, *Cs* out competes *Pp*, especially when *Cs* has relatively high density (see section 7.3.1). Therefore, in this model, we consider that only if *Cs* exceeds a certain density (e.g. 60% occupation), it can influence *Pp* life ability and growth rate.

The impact of grazers to these two types of plants is simplified to three effects: (i) slow down the growth rate, (ii) increase the mortality and (iii) help in seed propagation to further away areas than the neighbouring area of the plants, which is very difficult to quantify. Therefore, these terms are treated as small perturbations in the related processes.

*Spatial extension*
When the cell density becomes too high, the plant will extend to more distant areas beyond the finite number of discrete cells that are making up

the model. In the MAS model produced in this research, this is implemented by using Cellular Automation concept. Each cell having relatively lower plant density than some of its neighbouring cells with relatively higher density, will get a portion of the neighbouring regions plants, which is done by adding a small random number $P$ as the probability of extension in the calculation of the density variations. This is modelled for each plant based on local plant density and total density of plants. However, if the central cell is quite crowded (e.g. more than 50% covered by a certain plant) and at the same time, the neighbouring cells are quite occupied (e.g. more than half of the neighbours having more than 50% $Cs$ or $Pp$), then the plant growth will be constrained or even reduced (as part of mortality besides the seasonal and temperature influences). Plants then grow with reduced life ability and plant densities will be reduced by a small random portion.

*Seed dispersal*
Seeds disperse only when the plants reach a certain age (e.g. after reaching 50% and before 75% of its maximum life span). There are three ways for $Cs$ and $Pp$ to disperse their seeds (Eq. 7.6): (i) to the cell itself as a function of existing plant density $D_b$ and plant current $age$; (ii) to the neighbouring cells based on flow conditions (different spreading probability can be given according to flow velocity $v$) with a random probability $P_1$ due to the uncertainty of seed dispersal to neighbouring regions; and (iii) randomly taken by birds to other areas of the entire lake. For $Pp$, it hardly reaches more distant areas than its own rooting area and nearest neighbouring areas. Therefore, this MAS model considered mainly the roots for population reproduction for $Pp$, with a very small part of seeding by seeds which have less than 6% germination, as well as a very small random seeding by birds or ducks ($R$).

$$f(D, v, age, age_{(i,j)}, age_{(i-1,j)}, age_{(i,j-1)}, age_{(i+1,j)}, age_{(i,j+1)}, t, P_1, P_2, R) \quad (7.6)$$

### Model initialization
Initial density maps for both plants $Pp$ and $Cs$ are taken from GIS density maps in 1994 (Figure 3(a) and Figure 3(c)). Since the model starts to compute in April, in this case, the initial life stage is given by Eq. 7.7, and the minimal lifespan (*MinLifeSpan*) is 0 days for both $Pp$ and $Cs$. *StepToLive* is introduced for the number of days left for certain plants to stay alive. $R$ is taken as a random factor to distribute the plants in different life stages as the initial condition for the model, leading to the following expression:

*StepToLive = R * (MaxLifeSpan/4 – MinLifeSpan) + MinLifeSpan*  (7.7)

### 7.3.3 Sensitivity analysis

Three different sensitivity analyses were conducted in this case study: (i) sensitivity of environmental factors, (ii) of different processes and (iii) of initial conditions. In order to eliminate the influence from randomness added into the system, a fixed random number matrix is used and zero random seeding is considered for the sensitivity analysis of factors and processes, whereas the random number is not fixed for exploring the sensitivity of initial conditions, since the initial spatial patterns are generated and distributed randomly over the whole domain.

*Sensitivity of factors*
In this specific case, in order to select proper thresholds for certain factors such as water depth and velocity, some scenarios with ±20% and ±50% of certain values of above factors are implemented. Since there is lack of spatial distributed light information, the sensitivity of light condition is indirectly reflected by water depth since deeper areas are normally darker and more difficult for submerged plants to grow.

The sensitivity of water depth threshold and *x* direction velocity magnitude can be seen in Figure 7-8. Compared to velocity, the resulting occupation percentage is more sensitive to water depth changes. It also means that the light condition is more important for aquatic plant growth in this lake than flow dynamics.

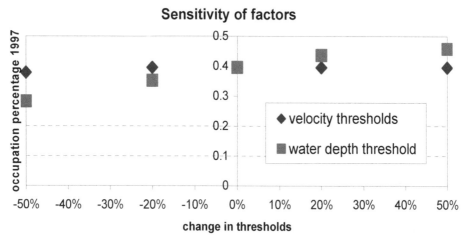

Figure 7-8 Sensitivity of external factors: water depth and velocity

*Sensitivity of processes*
The importance of different processes is also tested by removing one process each time. Resulting occupation percentages from 1994 to 1999 for different processes are shown in Figure 7-9.

From the sensitivity analysis of different processes in Figure 7-9, the most important process for Cs to colonize the lake is seen to be the seed dispersal process and growth process. Without the above two processes, Cs cannot colonize the lake. If only growth in the cells is considered, without spatial extensions and seeding as well as other processes like mortality, the occupation percentage does not increase. Scenario with "no interaction" obviously overestimated the resulting occupation percentages. It showed that the consideration of species interaction is also very important since different species constrain each other and compete for resources. Scenario with "no spatial extension" leads to the underestimation of the original modelling results.

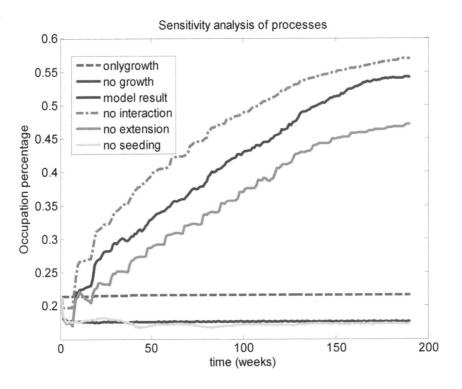

Figure 7-9 Sensitivity of processes

### *Sensitivity analysis of initial conditions*

For rooted plants, the initial seed bank is very important for determining the short-term spatial pattern development in an aquatic ecosystem. However, for long-term population colonization, the suitable habitat is more important. In order to check the influences from different initial conditions, a few scenarios are developed based on randomly generated seeds creating initial plant coverage ranging from 2% to 45% in Lake Veluwe. In doing so, the initial seeds can be anywhere in the Lake regardless water depth and flow conditions. The resulting plant occupation percentages are shown in Figure 7-10.

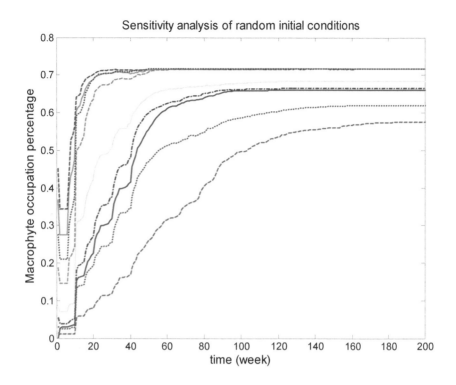

Figure 7-10 Occupation percentages with different initial conditions

The lines shown in Figure 7-10 are the resulting occupation percentages of *Cs*. It can be seen that after long time of running the model, the resulting occupation percentages become stable and exhibit less influence from initial conditions. Still, higher random initial coverage leads to higher probabilities for seeds to grow in areas that are difficult for mature plants to reach by seed dispersal processes, therefore leading to higher probability to have more spatial coverage. The sudden drop in occupation percentages right after the model begins to run is due to the super-imposed thresholds from environmental factors such as water depth and velocity. Such thresholds eliminated immediately the seeds that randomly located on the "wrong" places where plants cannot grow.

### 7.3.4  Analysis of model results

*Plant occupation percentages in Lake Veluwe*
An ensemble of model runs was conducted from year 1994 to 2001 (Figure 7-11). All model runs exhibit an increasing trend for *Cs* and decreasing trend for *Pp*. The results are slightly different due to the randomness added into the model. From the results shown in Figure 7-11, the occupation percentages of *Cs* seem to increase and exhibit a trend in colonization since 1994 to 1997 and afterwards towards a relatively stable status, whereas

densities for plant *Pp* decreased. Compared to the occupation percentages calculated based on GIS density maps, model results can well capture the increasing trend for *Cs* and the decreasing trend for Pp. It also shows a large difference between populations of *Cs* and *Pp*. During the first three years, the model well estimated the *Cs* occupation area but reached the saturation percentage earlier than the GIS map. Afterwards, the model results underestimated the occupation percentages mainly due to the simplified threshold constraints from water depth. The occupation percentage for *Pp* did not change too much and the modelled results well captured its trend in the model-running period, albeit with a slight under-estimation. There is a large decrease of *Cs* occupation in the year 2003, which is captured by the model.

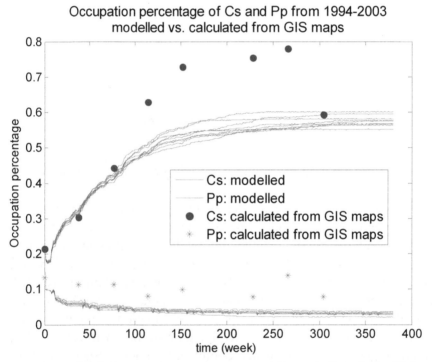

Figure 7-11 Occupation percentages of *Pp* and *Cs* in Lake Veluwe:
Modelled vs. calculated from GIS density maps

### *Resulting spatial patterns of macrophytes*

From the initially small occupation area to the large area of colonization, *Cs* became the dominant macrophyte in Lake Veluwe, which is well captured by the MAS model (Figure 7-12). By comparing the modelling results shown in Figure 7-12 (c) and Figure 7-12 (e) with the GIS density maps in Figure 7-6 (c) and Figure 7-6 (e), similar spatial patterns are captured by the MAS model. It shows that MAS rules developed in this study are quite reasonable for Lake Veluwe to simulate macrophytes' growth and spatial pattern dynamics. There are plants existing on some relatively small areas in the centre of the lake in the GIS map, but not in the model results, which is probably due to the seed dispersal effects by birds or ducks. Such effects are purely random phenomena that are very hard to quantify and are very difficult to be captured in any model.

The trend in the coexistence of *Pp* and *Cs* is well captured by the model as well. The coexistence of *Pp* and *Cs* dramatically reduced during the 5 years from 1994 (Figure 7-13 (a)) to 1999 (Figure 7-13 (b)), which can also be seen in Figure 7-6 (b), Figure 7-6 (d) and Figure 7-6 (f) in the GIS maps. In Lake Veluwe, the coexistence of two plants in the same area is now hardly observed, which is in accordance with the GIS maps.

Besides the occupation percentage, we have analyzed the quantitative differences in spatial patterns exhibited in GIS maps and modelled maps. The measures for the accuracy of modelling results for spatial pattern dynamics are the Probability of Detection (*PoD*) (Stanski et al., 1989; Werner et al., 2005) and the Threat Score (TS, or Critical Success Index (CSI)) (Schaefer, 1990). The *PoD* is calculated as shown in Eq. 7.6 and the TS (CSI) formulation is shown in Eq. 7.7. *F* is the calculated *PoD*, and *TS* is the calculated threat score, *A* is the total number of cells for the area correctly predicted compared to GIS map (*hits*), *B* is the number of cells predicted as water but that do have plants in the GIS maps (*under-estimated cells / missed*), and *C* is the total number of cells predicted as plants but that have no plants in the GIS map (*over-estimated cells / false alarm*). Figure 7-14 shows the map indicating the areas of *A*, *B*, and *C* for *Cs* in year 1997: the dark red area shows the correctly predicted cells, orange areas show under-predicted cells and dark grey with cross markers show the over-predicted cells.

Probability of Detection: $F = A/(A+B)$          (7.6)

Threat Score: $TS = A/(A+B+C)$          (7.7)

*PoD* is used to show what fraction of the observed plant coverage pattern has been detected by the model. It ignored the over-estimated spatial coverage predicted by the model. For example, *PoD* (1997) = 0.7841, which showed that roughly 78% of the observed pattern were correctly predicted by the MAS model. *TS* indicates how well the modelled "with plants" spatial pattern corresponds to the observed "with plants" spatial pattern, which considers the influence from both over-estimated (*false alarm*) cells and under-estimated (*missed*) cells. The bigger the difference between *PoD* and *TS*, the higher the false alarm rate becomes.

From Table 7-4, we can see that most of the years, MAS can predict *Cs* spatial patterns with considerably high accuracy. Compared to Figure 7-11, for year 1999, 2000 and 2001, the model under-estimated the plant occupation percentage, despite high *PoD* and *TS* values. The small difference between *PoD* and *Cs* indicates that in these three years there was very low possibility of over-estimation. Therefore, it is important in this case to combine *PoD*, *TS* and occupation percentage for result analysis.

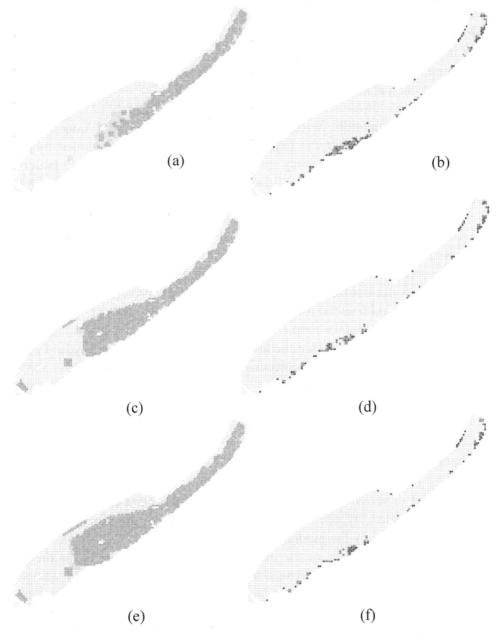

(a)                                    (b)

(c)                                    (d)

(e)                                    (f)

Figure 7-12 Modelled plant spatial patterns in year 1995, 1997 and 1999 (a) *Cs* (1995), (b)*Pp* (1995), (c) *Cs* (1997), (d) *Pp* (1997),(e) *Cs* (1999), (f) *Pp* (1999)

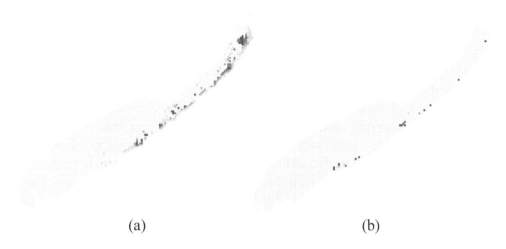

(a)                                    (b)

Figure 7-13 Modelled the coexistence patterns of *Pp* and *Cs* in year 1994 (a) and 1999 (b) (orange areas: coexistence of *Pp* and *Cs*)

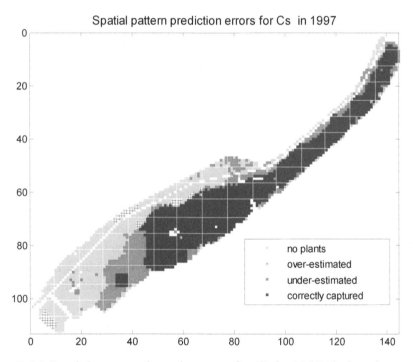

Figure 7-14 Spatial pattern detection map for *Cs* in 1997(dark red: correctly predicted; orange: under-predicted; black cross markers: over-predicted)

Table 7-4 Probability of Detection (PoD) and Threat Score (TS) for the modelled spatial pattern

| Year | 1995 | 1996 | 1997 | 1999 | 2000 | 2001 | 2003 |
|---|---|---|---|---|---|---|---|
| TS (*Cs*) | 0.6354 | 0.5889 | 0.7506 | 0.7300 | 0.7245 | 0.7291 | 0.7456 |
| PoD (*Cs*) | 0.7672 | 0.6542 | 0.7841 | 0.7562 | 0.7504 | 0.7437 | 0.8755 |

### 7.3.4  Discussion

This study explored the use of a Multi-Agent Systems (MAS) approach for modelling aquatic population dynamics, taking into account the simulation of growth and decay of two different macrophytes, and the interactions among macrophytes, grazers, and their living environment. During this research, a conceptual model was built using a MAS approach. The model coupled different processes into spatial patterns, and simulated macrophytes' population dynamics. Through both the conceptual model and a sensitivity analysis, dominant factors were investigated and MAS rules were developed. This application for Lake Veluwe showed the development of spatial patterns that very well captured the measured patterns contained in the GIS maps. Hence Multi-Agent Systems seem to be applicable for aquatic plant dynamics modelling provided proper ecological knowledge is taken into account. It holds the potential of representing a new working environment by coupling advanced modelling tools within a multi-agent system concept.

The patterns and aquatic plant spatial occupation are the results of multiple factors and processes. In this case, the small portion of seeds propagated to further areas implemented by randomness leads to the continuous spatial pattern to emerge, which showed the importance of seed propagation as dispersal process in spatial pattern development as well as the applicability of including stochasticity into the system. However, the randomness and the degree of randomness to different processes are hard to quantify.

Mathematical formulations and logical rules used in this study are in general simplified assumptions based on literature and biological and ecological knowledge. Additional studies need to be done to further improve MAS modelling with a better representation of each process involved. Furthermore, the feedback from plant growth to the change in flow pattern was not considered in this study, but is a very important aspect to be considered in future work.

## 7.4  Summary

This chapter gave an overview of a Multi-Agent System concept in which a summary of literature and properties was given. One case study was carried out to explore the use of a MAS approach in revealing the spatial pattern dynamics in aquatic ecological modelling, which considered the simulation of two different macrophytes, their interactions with their living environment. This case study proved that Multi-Agent Systems can capture the trend of plant growth development provided proper ecological knowledge is supplied. Such type of MAS model was rarely found in the literature for aquatic population spatial pattern dynamics modelling, however is proved in this case to be able to capture the development of spatial patterns, especially when the hydrodynamics is of relatively lesser importance compared to biological and ecological processes. Therefore, MAS holds great potential for simulating spatial pattern dynamics of aquatic populations. Compared to cellular automata based models, MAS shows more flexibility and a better representation of reality in practical applications.

# Chapter 8

# A synthesis of physically-based water quality modelling and multi-agent-based population dynamics modelling

## 8.1 Introduction

The previous two chapters discussed the use of discrete Cellular Automata (CA) and Multi-Agent System (MAS) on modelling spatial pattern dynamics of aquatic plants under simplified hydrodynamics and water quality conditions. However, spatial pattern dynamics for aquatic populations are often spontaneous results of the dynamics of the flow, the changes of substances in water bodies as well as the dynamics of plant biological growth, etc. Therefore, it seems relevant to explore a combination of these two different approaches towards aquatic plant spatial pattern dynamics, into one simultaneous modelling framework.

Hydrodynamics and water quality processes can be well modelled by continuous differential equations, which form the basis of physically-based water quality modelling package such as the Delft3D software suite. Meanwhile, the previous chapter showed the potential of using MAS for simulating spatial pattern dynamics in aquatic ecosystem modelling. Therefore, this chapter aims to show one way of developing a modelling framework to link continuous processes such as hydrodynamics and water quality with discrete processes such as spatial pattern evolution of aquatic plants by demonstrating a synthesis of a physically-based model with a multi-agent-system model. Section 8.2 gives a short introduction of the background knowledge concerning multi-processes in representing spatial pattern dynamics in aquatic ecosystem modelling. It also describes how to include external modules in the Open Process Library of the Delft3D-WAQ package. The processes considered in the synthesis model are represented in section 8.3. A test case is developed for a sensitivity analysis of different factors and processes involved and the application of this synthesis model is implemented for Lake Veluwe, which is described in section 8.4. Section 8.5 shows the results of the sensitivity analysis and of the simulation in Lake Veluwe. A summary of this chapter is given in section 8.6.

## 8.2   Delft3D-WAQ open process library

As described in chapter 2, aquatic population dynamics is influenced by its own properties, environmental factors (e.g., water motion, nutrient availability, sediment transport), as well as meteorological conditions (e.g. temperature and light). At the same time, they provide feedback to the environment they live in (e.g. through changes in flow patterns and nutrient dynamics).

However, technically it is not an easy task to implement such approach into a synthetic modelling framework. It involves the coupling of different processes with different spatio-temporal scales, and the integration of different types of information and data sources, as well as the linking of different modelling concepts. In terms of appropriate scales, advection-diffusion processes can have the dynamics of seconds, while water quality conditions can have a characteristic time of hours or a day, whereas aquatic plants generally change much slower with a characteristic time of about one week or even longer. Although integration in aquatic ecosystem modelling is not novel (see chapter 3), a synthesis of multi-agent based discrete modelling with physically-based continuous modelling approaches for simulating aquatic population dynamics has not yet been found in references.

### 8.2.1   Background

A first version of the water quality module in the Delft-3D software package was developed in the 1980s and the DELWAQ process library inside water quality module was initiated since the early 1990s (WL | Delft Hydraulics, 2006b; WL | Delft Hydraulics, 2007). The DELWAQ process library follows an object-oriented approach where each substance is an object. The interaction between different substances as water quality kinetics (called processes) is also seen as an object. This has very powerful implications. In the Open Process Library, there are several water quality components: substances (e.g. concentrations or density of certain nutrients), processes (e.g. growth of algae), items (e.g. input/output variables of a process) and fluxes (a special class of items consists of fluxes between substances). The library currently includes more than 200 substances/organisms and fractions of substances which are relevant, as well as more than 400 processes that represent the interactions between substances. Besides, some library functions can be re-used for different substances having the same behaviour, by changing parameters or inputs into the function. Still, not all processes and substances can be included into this process library, especially the processes that represent the spatial pattern dynamics of macrophytes and other populations that are different from phytoplankton. Therefore, it is important to have an open interface which allows users to add substances and processes, as required in specific projects.

The concept of an Open Process Library (OPL) was developed precisely for such use. The DELWAQ process library is an extensible library of water quality components. The OPL is the developer environment of the existing Delft Process Library Configuration Tool. Such developer environment is to facilitate the process of "creation" of new substances, processes, parameters and their settings to be part of the whole DELWAQ Library. Thus by using the Open Process Library and a FORTRAN compiler, substances, processes and parameters can be added into the existing library. Once the substances, processes, etc. are added, the users can easily decide to switch them on or off without having to bother about the software connections between them. Besides, users can also specify which items they want to be editable through the user interface, or which items they want to add to the output list. After specifying the water quality components, the model itself deals with these components. Furthermore, in this open process library, the system takes care of all required connections, such as invoking component processes, outputs and fluxes between the substances. When one switches on certain process, the processes supply inputs to be automatically switched on. The system identifies for all editable items whether they are provided as a constant, as a single time function for the whole area, as a spatially distributed constant or as a spatially distributed time function. The resulting model configuration of the water quality model contains all choices made and is saved for later reference purposes.

## 8.2.2 Adding modules into the DELWAQ open process library

The procedure for adding modules (e.g. the MAS aquatic plant dynamics module) into the DELWAQ OPL is briefly outlined below (WL | Delft Hydraulics, 2006b):

- Creating a substance group and substances
A substance group is the group name of several different substances, in this case, "macrophytes", which is to calculate the processes related to macrophytes. The substances here can be the different types of macrophytes the developer would like to add into this substance group. Each substance has its specific properties, which are defined here as variables.

- Creating processes
With the new substances recently created, the next step is to create processes that describe the interactions between substances through the variables of each substance. Each process has a unique name and ID, and the developer needs to define the inputs, outputs (named items in the

process library) and fluxes of the process. The fluxes are special items that affect the substances. The developer needs to define where the fluxes act on and the factors with which they act.

- Algorithm development (coding)

The DELWAQ open process library facilitates the code generation by generating a skeleton of the FORTRAN code. Inside this skeleton, the locations for adding algorithms for processes are given that allow the user to include mathematical and logical descriptions. In such an environment, the dynamic changes of substances or variables representing substance properties can be defined as fluxes. Besides, due to the special consideration of spatial extension of aquatic plants, values in the neighbouring cells are needed to be read and then added into functions or conditions that can be implemented in such a framework as well. Again, the proper scales and units need to be specified correctly.

- *Dll* generation and process installation

A dynamic link library can be generated and then installed into the DELWAQ open process library. Then, a process library that includes newly specified substance groups is generated and ready for use.

## 8.3 Synthesis of multi-agent based aquatic plant dynamics model into open process library

### 8.3.1 Purpose

The main purpose of a synthesis of modelling paradigms is to have an online coupling of different dynamical processes involved in aquatic ecosystem modelling- in this case, macrophytes' dynamics modelling (two different types of sub-merged aquatic plants: *Potamogeton pectinatus* (*Pp*) and *Chara aspera* (*Cs*)) are taken as examples. It includes the dynamical changes of flow patterns and water quality factors to see how aquatic plant patterns can be influenced. In this synthesis of models, it can include information and data by modelling intermediate results as inputs for other processes, which sometimes cannot obtain from measurements.

### 8.3.2 State variables and scales

Concerning macrophytes, in this case submerged macrophytes, many different properties need to be presented in the modelling process and results. Given the large number of macrophytes in a small area (e.g. *chara* can have 1000 stems per $m^2$), we consider one of the state variables of vegetation is denoted as *stem density* ($Np$, stem/$m^2$, the number of stems per square meter), which was represented in the previous chapter as a percentage of cell occupation with a multiplier of 100. The increment of *stem density* per time step is assumed to be modelled based on a logistic function as also described in the previous chapter. In the computational segments, there is a maximum stem density for each species as carrying capacity defined as *Np_max* (stem/m$^2$). Consequently, the *Np/Np_max* <=100%.

Each stem has some properties, such as *age*, *lifespan* (how long a macrophyte can live), *height* (height from bed level to the top of the macrophyte), and *diameter*, etc. Due to high density of macrophytes, populations of macrophytes located within one computational cell are considered as one agent and their values are assigned at the centre of each computational cell. Thus the model can be seen as a meta-population model or super-individual (Grimm and Railsback, 2005) model. The precaution is needed for the units of each state variable in the programming process. This is determined by how the program couples the different scales of the hydrodynamics model, water quality model with macrophytes' growth model. Assuming a unified spatial scale for all processes, only the time steps of different processes need to be coupled. For water quality processes, a default time step has the unit of *one day*. Therefore, the weekly calculation in the previous chapter is modified to be as *DELT* day where *DELT* is the time step of the water quality module with the unit of *day*.

Besides the variables related to macrophytes, there are many other variables related to hydrodynamics and water quality processes as environmental factors that have to be added into the modelling system, in this case, water depth, flow velocity, extinction coefficient, wind, etc.

### 8.3.3 Processes

The synthesis model consists of several processes as shown in Figure 8-1, and among them, the aquatic plant dynamics module includes germination, growth, spatial extension, competition, mortality, seed dispersal. These processes are interacting with the processes of hydrodynamics and water quality. As indicated in the figure, the latter two processes are well developed and modelled by the Delft3D software package, whereas, the processes of aquatic plant dynamics can be modelled by using a multi-agent system concept as described in Chapter 7.

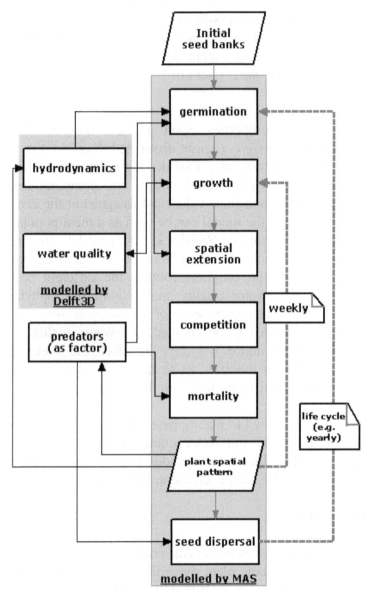

Figure 8-1 Processes involved in aquatic plant dynamics modelling

Hydrodynamic conditions can be treated as engines for seeds transportation and for spatial extension, and also can be a limitation for plant germination (higher flow leads to lower germination). Meanwhile, the resulting spatial pattern also can have very important impacts on hydrodynamics patterns although this is not the main focus of this research. In terms of water quality processes, we see that the most direct link should be that the growth of plants requires nutrients from its local area including the waterbed and water column, and the photosynthesis of plants produces oxygen into the water body. This can influence the nutrient level and lead to a better water quality

(Asaeda et al., 2000) in terms of reducing excessive nutrients from the water column.

## 8.3.4  Sub modules and modelling procedure

By using the open process library in the water quality module of Delft3D software package, the aquatic plant dynamics module is embedded into the water quality module and is calculated together with other water quality processes when the model scenarios are setup. As introduced previously, the aquatic plant dynamics module includes many sub modules and they are implemented as outlined in Figure 8-2. The substances group in this case is "macrophytes" added into the DELWAQ process library and two different types of sub-merged aquatic plants are considered as example: *Potamogeton pectinatus* (*Pp*) and *Chara aspera* (*Cs*)). The processes included here are based on the detailed description given in Section 7.3.2 with minor revisions to fulfil the requirements in the Delft3D software package.

The "macrophytes" module consists of five processes, namely: growth, spatial extension, interaction/competition, seed dispersal and mortality. A brief introduction of each process is given here for most of the processes, while a more detailed explanation is given in Chapter 7 (mainly in Section 7.3.2).

*Growth* is a process having density as a value to represent growth under the conditions that temperature and flow conditions are favourable, which is modelled with a logistic function. The density flux in this process includes both a deterministic part represented by a logistic function, and functions or rules for other processes as well as a random portion that includes certain stochastic variations. In terms of the influences from non-favourable environmental conditions, some thresholds are adopted for the influences of flow and light conditions. The growth process is implemented separately for each species.

*Spatial extension* acts as a diffusive process that occurs when a certain area is having a relatively high density of e.g. aquatic plants that need neighbouring areas as growth space. This is implemented as one process for both species due to the occupation of space by both plants. Two sub-processes are considered: (i) *too crowded* with both agent and neighbouring agents having quite high densities (ii) *extendable* where neighbouring agents have high densities but the agent itself has relatively low density. Randomness is easily added into the diffusion processes.

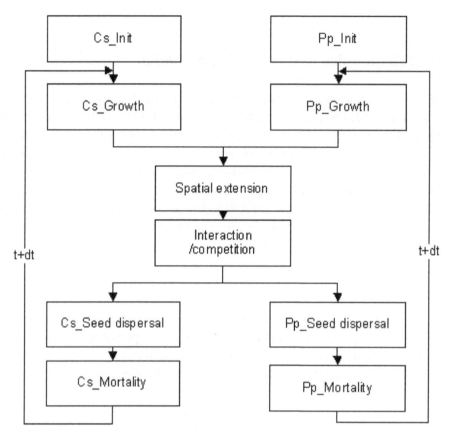

Figure 8-2 Procedure of process implementation for aquatic plant dynamics module

***Interaction/competition*** is between two species. Therefore, it is also one process for both species. Such interaction occurs within a computational cell and the functions and rules are the same as described in chapter 7.

***Seed dispersal*** is one of the main processes that keeps the system developed and allows it to regenerate for multiple years. It is a species-specific process due to the large differences among macrophyte species. It happens only when macrophytes become mature. In this case we assume that seed production happens when macrophytes reach certain age, e.g., older than 50% and younger than 75% of their maximum age.

Flow condition is an important factor for seed dispersal of *Cs* because a large number of the tiny seeds produced by *Cs* are carried by the flow. Besides, high randomness is involved in such process, especially for *Cs*. Therefore, a random number is added into this sub-module. Seeds travelling distance is based on flow velocity. Since *Pp* is mainly reproduced by its potato-like roots, it regenerates mainly based on previous year's seed bank.

***Mortality*** means that when aquatic plants reach their life span then they will die naturally.

## 8.4 Simulation experiments

A sensitivity analysis is conducted and the application to Lake Veluwe was explored in this research. The process of simulation experiments includes the schematization of the model domain, model configurations, processes, input data and information, as well as the construction of different scenarios.

Model schematization and initialization are for the modules of hydrodynamics, water quality and macrophytes growth. In this case, spatial steps are selected once for all the processes and modules, therefore, only one schematization is realized and used in the whole modelling procedure. In the Delft3D software package, computational grids can be rectangular or curve linear. Nutrients concentrations in the research area are assumed to be non-limiting for aquatic plant growth. In order to calculate spatial pattern dynamics of aquatic plants, the initialization of macrophytes spatially distributed density maps are needed.

### 8.4.1 Sensitivity analysis

*Schematization and flow configuration*
A sensitivity analysis is carried out for some factors and processes included in the synthesis model using several test cases composed of simple rectangular domains with uniform bathymetry and cell size of 100 by 100 m$^2$. Figure 8-3 shows the schematization of the computational domain.

A very simple flow field is modelled for this test case. It has a constant flow rate coming from the left boundary into the modelling domain, while a zero water level is considered as the right boundary. Such boundary condition creates a constant velocity of about 0.14 m/s and a constant water depth of around 1.5 meters is considered. In such case, the influence of water depth and velocity is homogeneous in the whole domain, and therefore does not influence the spatial distribution of the aquatic plant species.

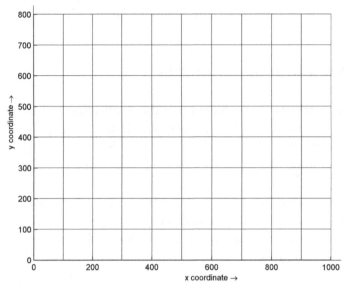

Figure 8-3 Schematization example for sensitivity analysis ($100 \times 100$ m$^2$)

### *Initial aquatic plant maps*
Initial density maps for *Pp* and *Cs* are shown in Figure 8-4 separately. Such initial maps are considered the same for all the testing scenarios.

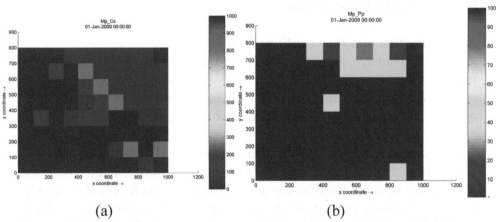

(a)                                                          (b)

Figure 8-4 Initial density maps for sensitivity analysis: (a) *Cs* and (b) *Pp*

### *Sensitivity analysis of factors and processes*
Many different processes and factors are included in the construction of the substance group "macrophytes" although the degrees of importance of them are different. Before using this substance group for spatial pattern dynamics simulation, a sensitivity analysis of different processes is needed. In this case, this is achieved by trimming of processes (e.g. spatial extension, seed dispersal, interaction) which is similar as the sensitivity analysis in Chapter 7.

### 8.4.2  Application in Lake Veluwe

***Schematization***

In this synthesized model, a coarse rectangular modelling grid with the size of about 100 by 100 square meters is adopted (Figure 8-5). The bathymetry of the Lake Veluwe, as shown in Figure 8-6, was supplied by Dutch water board.

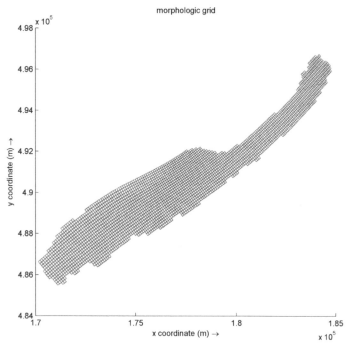

Figure 8-5 Lake Veluwe schematization

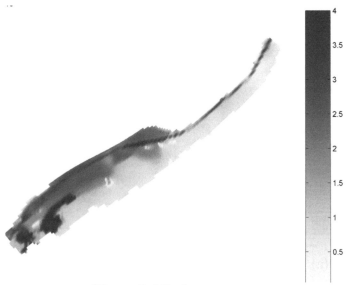

Figure 8-6 Bathymetry map

### Initial plant maps

Initial plant maps obtained from GIS density maps of the year 1994 were shown in the previous chapter (Figure 7-6 (a) (b)).

### Simulation procedure

For the application in Lake Veluwe, a simplified flow dynamics driven by wind is adopted and system does not consider nutrients limitation due to the aquatic plant ability of absorbing nutrients from the lakebed, therefore it is assumed that nutrients are sufficient even without loading from outside of the domain. In this application, water depth and velocity are with dynamic changes, especially velocity due to the changes in wind condition.

## 8.5   Analysis of results

### 8.5.1  Sensitivity analysis results

Several scenarios are produced based on the trimming of processes from general considerations (Table 8-1). Resulting density maps are shown in Figure 8-8. These results are compared with the scenario considering all the factors and processes shown in Figure 8-7.

Table 8-1 Sensitivity analysis scenarios

| Scenarios | Processes |
|---|---|
| S1 | general configuration (with all processes) |
| S2 | no spatial extension, with growth, interaction, and seed dispersal and germination |
| S3 | no spatial extension and no species interaction, with growth and seed dispersal and germination |
| S4 | only growth |

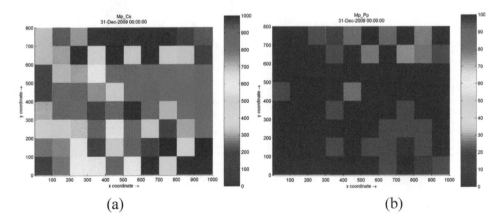

(a)                                                                (b)

Figure 8-7 Results of scenario 1 with general configuration (a) *Cs*, (b) *Pp*

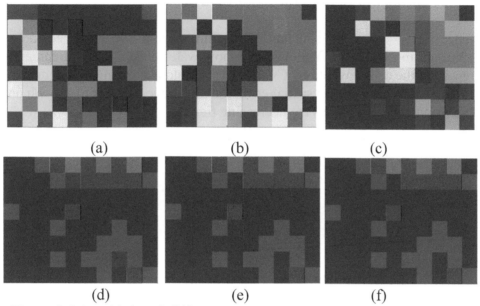

Figure 8-8 Sensitivity of different processes, for *Cs*: (a) S2 (b) S3, (c) S4; for *Pp*: (d) S2 (e) S3 (f) S4

From above figures, we can see that different processes have different influences to the spatial pattern dynamics of *Pp* and *Cs*. For *Pp*, the spatial patterns of *Pp* resulting from S1 (Figure 8-7(b)), S2 and S3 (Figure 8-8 (d) and (e)) are almost the same indicating that the inclusion of spatial extension and growth can already represent *Pp* spatial pattern quite well, and the inclusion of other processes do not obviously influence the results. For the spatial pattern dynamics of *Cs*, it seems that all the processes are important. S2 (Figure 8-8 (a)) did not consider spatial extension showing obvious different spatial pattern compared to S1 (Figure 8-7(a)). Besides, compared to S3 (Figure 8-8 (b)), S1 has less spatial coverage due to the inclusion of the interaction between *Pp* and *Cs*. S3 did not consider interactions between *Pp* and *Cs,* therefore *Pp* does not limit the growth of *Cs*, therefore S3 resulting spatial map with higher density compared to S1. For the scenarios S4, only the cells that do not reach their maximum life span continue to grow and no newly grown stems are added into the map (Figure 8-8 (c) and (f)). Besides, due to the decay processes are not included in S4, after reaching maximum life span, the spatial pattern remains as it is till the final computation.

In general, the sensitivity analysis shows that the spatial extension and seed dispersal and germination are relatively more important for aquatic plant spatial extension, which is similar to the results obtained in chapter 7. In this way, biological/ecological diffusive processes are included into the model. Furthermore, the interactions between species are seen to be indispensable.

## 8.5.2  Modelling results for Lake Veluwe

After a few years of simulation, resulting spatial patterns become as indicated in Figure 8-9. Yellow areas show water coverage and green areas represent the cells covered by *Cs* with darker colour showing higher density. *Cs* has extended to larger spatial coverage after several years of growth and extension, whereas *Pp* has the decreasing trend.

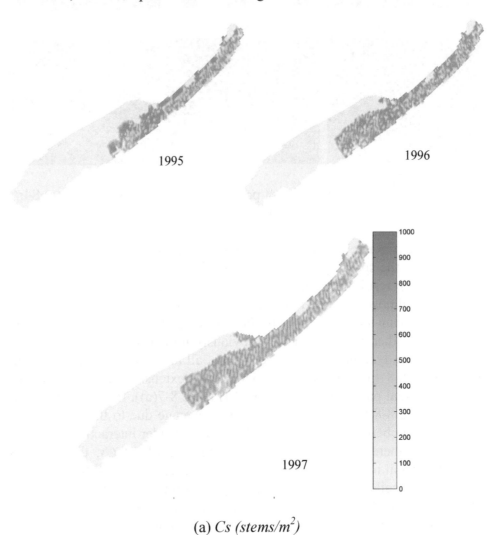

(a) *Cs (stems/m²)*

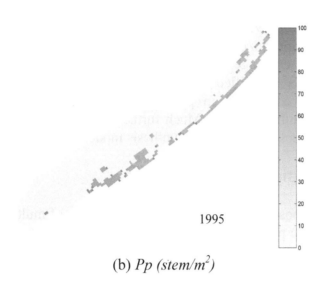

1995

(b) *Pp (stem/m²)*

Figure 8-9 Resulting density maps (a) *Cs*, (b) *Pp*

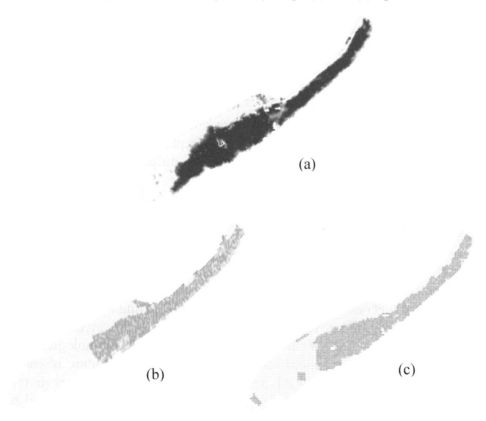

Figure 8-10 Resulting density maps (*Cs*, 1997) of (a) GIS density map (b)
synthesis model, and (c) MAS in Chapter 7

Comparing to the GIS density map and resulting *Cs* map from MAS in Chapter 7 for year 1997 (Figure 8-10), the resulting density map from synthesis model appears quite similar coverage. One of the possible reasons of slightly less coverage compared to previous chapter may be that hydrodynamics is simplified to a wind-driven flow field. Besides, the random seeding considered in previous chapter is not considered as a process in the synthesis model. Much further sensitivity analysis and better model development are needed for synthesis modelling approach.

## 8.6    Discussion and summary

This chapter proposed a way of achieving a synthesis of multi-processes for non-linear aquatic population dynamics modelling. It is based on coupling a continuous Physically-based water quality model (DELFT3D-WAQ) and a discrete multi-agent macrophytes' growth model. A sensitivity analysis was conducted which showed that besides plant growth, the processes of seed dispersal, spatial extension and species interactions are indispensable processes in modelling aquatic plant spatial pattern dynamics, especially for the extension of patterns representing biological diffusive phenomena. In the case study of Lake Veluwe, the synthesis model produced similar results as obtained in chapter 7. The implementation of the synthesis of this multi-process, non-linear system shows a possible way of integrating continuous processes with discrete processes. Moreover, this synthesis approach includes biological/ ecological growth and diffusive processes, as well as local effects in population growth into a coherent modelling framework and therefore further enhances the conventional differential-equation based modelling.

The above example shows a challenging field which calls for further research on mechanism understanding and mathematical formulation of processes. This synthesis of models is just the first trial of integration of hydrodynamics and water quality continuous equations with a discrete MAS aquatic plant growth model. Although Van den Burg (Van den Berg, 1999) has indicated that one of the reasons for *Cs* out-competing *Pp* is the increase of pH in this lake, more research is still needed for further investigation of *Cs* colonization in Lake Veluwe. A better formulation of biological and ecological processes (e.g. decay process) is needed in the future research. The same applies to the feedback mechanisms from biotic processes (in this case plant growth) to the hydrodynamics of the surrounding flows and to morphological changes. This biogeomorphological research is becoming a separate field of science in itself (Uittenbogaard, 2003; Baptist, 2005; Paarlberg et al., 2005; Temmerman et al., 2007).

# Chapter 9

# Conclusions and recommendations

## 9.1 Conclusions

Spatial patterns evolving from aquatic population dynamics are vital for habitat sustainability and aquatic ecosystem health. Modelling such dynamics can be very helpful for managing the aquatic environment. This research focused on how to reveal the most important processes and factors involved in observed spatial pattern dynamics of aquatic populations and how to simulate such dynamics with various modelling paradigms and many different data sources.

This research is cross-disciplinary and involves expertise in hydrodynamics, water quality, and ecology as well as hydroinformatics modelling techniques. In this thesis, the main factors and processes involved in aquatic population dynamics are listed and existing modelling tools for analysis and simulations of spatial pattern dynamics are reviewed. Different modelling approaches with different emphases and different modelling requirements were explored and applied to real world situations in order to increase the understanding of the underlying mechanisms and processes involved. The thesis covers different case studies ranging from large scale North Sea algal population simulation to meta-population dynamics in inland lakes and to individual-oriented macrophytes' growth dynamics in a small pond. It combined many spatial data sources from remotely sensed imageries, GIS density maps, or high resolution photos. Based on such variety of data and expert knowledge, this research explored and developed alternative approaches for aquatic population spatial dynamics modelling. In doing so, differential equation based models were enhanced, discrete modelling techniques were extended, and Multi-Agent System (MAS) approaches were explored in a new area of aquatic population spatial pattern dynamics modelling. The importance of incorporating spatially heterogeneous information and local effects as well as ecological diffusive processes was demonstrated in a number of case studies developed in this thesis. A synthesis of multi-process non-linear ecohydraulics systems is proposed for utilizing the "best of all" of different types of modelling techniques in order to better represent the spatial pattern dynamics having different spatial scales and temporal characteristics. Such synthesis of continuous physically-based models with a discrete MAS population dynamics model was implemented in the Delft3D software package at Deltares, making it available to other users as well.

The relationship between aquatic populations and their physical environment, as well as the interactions with other living organisms were described as the domain knowledge for this research. The alternative approaches followed in this research proved that the spatial pattern dynamics of aquatic plants are the spontaneous result of multi-processes and multi-factors. Since not all processes and factors could be included in the modelling systems, a sensitivity analysis is often needed for selecting the dominant factors. Data-driven modelling techniques proved to be quite good for main factor selection and non-spatial population prediction, provided sufficient data is available.

The use of discrete modelling techniques and agent-based modelling approaches explored were seen to represent the spatial pattern dynamics quite well in a quantitative sense in this study, while also achieving a better understanding of some fundamental underlying mechanisms. The local effects and individual properties of aquatic plants are indispensable processes and factors in forming global spatial patterns. Besides correctly simulating the spatial pattern dynamics and achieving a better understanding of the phenomena involved, some of the applications also have the capability of more quantitatively predicting spatial pattern dynamics. The results of the various combined modelling approaches can further help in quantifying the spatial habitat complexity. Along with the fast development in measurement systems and the wider availability of spatially distributed data from sensor networks, radar observations, etc., the combination of multi-data sources and different modelling approaches holds great potential for better simulation and prediction of spatio-temporal aquatic population dynamics, and can contribute greatly to achieving better water management strategies and for sustainable development of the aquatic environment.

## 9.1.1  Differential equation-based approaches

Differential equation-based models are often derived from conservation principles for homogeneous conditions that can represent averaged behaviour and spatially smoothly changing phenomena quite well when the mechanisms are well understood. Some physically-based model implementations (e.g. Delft3D-BLOOM/GEM) couple algae species competition formulation with traditional differential equation-based models for flow and transport phenomena (Delft3D-FLOW/WAQ). However, we still do not have enough knowledge to represent all the main processes involved in aquatic ecosystem dynamics adequately. In a quite turbulent environment, hydrodynamics is likely the main dominant factor for spatial pattern dynamics; hence, differential equation-based models show good performance. In the example shown in Chapter 5, some of the data used in the original algal bloom model for the North Sea have relatively high uncertainty, but these data sets are very sensitive factors to correctly predicting the algal population dynamics. Therefore, this research addressed a possible way to enhance the current differential equation based model Delft3D-BLOOM/GEM with remotely sensed imageries as the replacement of those uncertain data. In this way, a better representation of spatial heterogeneity of phytoplankton dynamics in the North Sea was achieved. This pilot application gave an important message that the inclusion of spatially heterogeneous data leads to a relatively better representation of spatial pattern in the model. Therefore, one of the possible ways to enhance existing differential equation-based models is to include more spatial data. However, some of the processes related to ecological growth and diffusive phenomena cannot yet be included in conventional equation-based modelling approaches, and require alternative mathematical formulations.

## 9.1.2 Discrete cellular automata models

When the behaviour of individual plants and/or local effects are vital to the development of spatial patterns, discrete Cellular Automata (CA) and agent based modelling techniques seem more suitable for representing the spatial pattern dynamics than equation-based formulations. This research demonstrated that CA based models are good in representing discrete phenomena like plant growth which are known to depend on local effects. Since CA models need proper selection of rules, grid size and neighbourhood schemes, this study combined various data sources like time series of high resolution photos, biological knowledge from experts, and time series of meteorological data in the selection of proper grid size and neighbourhood schemes. The time evolution of spatial images proved to be very useful for simulating aquatic plant growth. It also proved that when the selected cell sizes correctly represent the biological characteristics of the modelled species, the growth processes of individual plants can be better represented. From this point of view, domain knowledge again becomes vital in achieving better modelling approaches and modelling results. Alternative techniques like data-driven techniques were applied in deriving the local rules for the CA models, which increased the flexibility by extending simple if-then rules to highly non-linear relations.

However, CA type modelling approaches are constrained by their own concept and computational cell configurations. In case interacting species coexist in the same local environment and competitions among them are within each computational cell, especially when such competitions are mainly due to species own properties, traditional CA becomes limited due to its lack of flexibility.

### 9.1.3  Multi-agent systems

In order to include better the individual/super-individual properties of particular species and account for the interactions among different species, the use of Multi-Agent Systems (MAS) was explored in this research by integrating different data (e.g. GIS data), different processes, and different techniques. MAS systems could capture the correct spatial development trend of plant growth, when proper ecological knowledge and data were provided. Such type of MAS models hold great potential in aquatic population dynamics modelling. Compared to traditional CA, MAS emphasizes the properties of the agents or individuals, and can include more processes that are realistic as well as being able to model more agents that are movable. It is much more flexible and more applicable in terms of a better representation of reality and practical uses, in particular when seed dispersal phenomena or large excursions of grazers are involved.

There are several vital aspects in MAS models. First, the most important aspect is the role of expert knowledge. Most of the rules and functions created in the model used here were derived from expert knowledge and past experiments and analyses. Secondly, due to the ubiquitous features of aquatic ecosystem being stochastic, nonlinear and complex, some processes, which are hardly quantified by any equation, e.g. seed dispersal, impact from birds and fish and the selection of the directions for spatial extension, were quantified as random terms in the model. Thirdly, spatial data from annual GIS density maps were used as initial conditions and calibration data for aquatic plant growth modelling. These data, together with the bathymetry and water depth information, became the most important spatial information to construct MAS model. Furthermore, the results from the MAS model showed that both the external environment (e.g. water depth and temperature) and individual properties of aquatic plants' species are the determining factors for the evolving spatial pattern dynamics.

### 9.1.4  A synthesis of a multi-process ecohydraulics system

The general features of aquatic ecosystem are related to multi-process, multi-scales, nonlinearities and randomness. There is the need to have a modelling framework for including both the turbulent environment and aquatic population dynamics. Such framework then becomes an interdisciplinary approach. Since the aquatic plants' spatial pattern dynamics are in part determined by local effects and species properties, but since they live in water bodies, hydrodynamics and water quality processes are also vital for the pattern formation. Such patterns in turn also have feedback on the flow and water quality. Therefore, a synthesis of multi-process, nonlinear feedback systems modelling was developed in this research. This synthesis included both continuous processes conventionally modelled by differential equations and discrete processes in this case modelled by Multi-Agent System concept.

The framework considers aquatic ecosystems in a holistic way, which incorporates many aspects of domain knowledge and technical implementations. With an object-oriented modelling framework, this synthesis was successfully implemented in the DELWAQ open process library of the Delft3D software package. Sensitivities of several factors and processes were tested with a simple test case and the synthesized model was applied to explore the spatial pattern dynamics in Lake Veluwe. The success of the synthesis of this multi-process, non-linear system gave a possible way for the integration of continuous processes with discrete processes. Moreover, the synthesis approach implemented in this research included biological/ecological growth and diffusive processes, local effects in population growth into the conventional modelling framework and therefore further enhanced differential-equation based modelling.

## 9.2 Recommendations

Clearer understanding of one system could lead to a better mathematical formulation in the modelling of the system with the inclusion of sufficient ecological knowledge. Modelling of aquatic ecosystems, however, is difficult due to the complexity involved and interdisciplinary characteristics of the system. Further research is needed on better mathematical formulations of biological/ecological processes in aquatic ecosystem modelling. Such kind of research needs more relevant domain knowledge, more spatially distributed information and data on different scales, as well as in situ measurements and laboratory experiments. The collaboration between modellers and specific domain experts was one of the determining factors for the development of the models in this research. It is very important to the success of choosing modelling approaches and to the better understanding of the real world complex problems. Collaborations between modellers and domain experts need to be further strengthened in future research.

More sensitivity analyses can be done in future work, not only for single process or factor, but also the interactions among different processes and factors reflecting the interactions of different species as well as the interactions of species and environment. Such kind of sensitivity analysis is very important for the analysis of biodiversity in aquatic ecosystems.

Besides, because the key to understanding complex systems often lies in understanding how processes on different scales and hierarchical levels are bound to each other, more sensitivity analyses are also needed to be carried out on different scales as well as for the selection of modelling scales in aquatic ecosystem modelling.

Uncertainties in models and data are unavoidable, but can be reduced under further studies including better mathematical formulations and sensitivity analysis. Besides, more measurement data and better interpretations of available spatial and temporal data can further reduce the uncertainties. Therefore, more measurements need to be carried out on different spatial and temporal scales that can reveal main processes and factors for representing spatial pattern dynamics at different scales. Still modelling is just one of the ways to obtain a better understanding of the phenomena and is sometimes far from really understanding what is going on in aquatic ecosystem. Therefore, modelling results need to be analyzed by combining specific domain knowledge from specialists with laboratory experiments, in situ measurements as well as results from other models, in order to have a better understanding of the real situations.

This research introduced a synthesis in the modelling of multi-process and non-linear ecohydraulics systems, which holds great potential of adequately representing features of nonlinearity, randomness and complexity in aquatic ecosystems. Much further research is needed to improve this synthesis model. One of the studies in the future research can be the inclusion of feedback mechanisms from the growth of aquatic populations to hydrodynamics and water quality. There are a number of references on modelling the impacts of aquatic plants on the hydrodynamics, and also some related to aquatic water quality. Further collaboration with specific domain experts, more sensitivity analyses, better calibration of the developed approaches, more spatio-temporal data acquisition, and better mathematical formulations of different processes, are needed to achieve a more matured modelling framework for aquatic ecosystem modelling. The explosion in sensor technologies will provide ever increasing volumes of data with unprecedented spatial and temporal resolution, which can stimulate further developments and extend of approaches introduced in this research, leading to new knowledge on modelling aquatic ecosystem dynamics.

Furthermore, modelling applications need to be extended to real applications in order to be used to analyze the impacts from man-made disturbances, natural disasters, global warming and effects of climate change to changes in aquatic ecosystems, to facilitate decision-making process for better aquatic environmental management.

# References

Abbott, M.B., 1979. Computational hydraulics: elements of the theory of free surface flows. Monographs and surveys in water resources engineering. The Pitman Press, Bath.

Abdullah, K. and et al, 2000. Remote sensing of total suspended solids in Penang coastal waters, Malaysia, Malaysian Government IRPA.

Alberti, M. et al., 2003. Integrating humans into ecology: opportunities and challenges for urban ecology. BioScience, 53(12): 1169-1179.

Almeida, C.M., Gleriani, J.M., Castejon, E.F. and Soares, B.S., 2008. Using neural networks and cellular automata for modelling intra-urban land-use dynamics. International Journal of Geographical Information Science, 22(9): 943-963.

Anderson, D.M., Andersen, P., Bricelj, V.M., Cullen, J.J. and Rensel, J.E., 2001. Monitoring and management strategies for harmful algal blooms in coastal waters, Asia Pacific Economic Program, Singapore, and Intergovernmental Oceanographic Commission Technical Series, Paris.

Arias, M., Li, H., Blauw, A., Peters, S. and Mynett, A.E., 2009. Enhancing Delft3D-BLOOM/GEM for algae spatial pattern analysis: Filling missing data in RS images, Proceedings of the International Conference on "Science and Information Technologies for Sustainable Management of Aquatic Ecosystems", the joint meeting of the 7th International Symposium on Ecohydraulics and the 8th International Conference on Hydroinformatics, Concepcion, Chile.

Asaeda, T., Trung, V.K. and Manatunge, J., 2000. Modeling the effects of macrophyte growth and decomposition on the nutrient budget in Shallow Lakes. Aquatic Botany, 68(3): 217-237.

Asaeda, T. and Van Bon, T., 1997. Modelling the effects of macrophytes on algal blooming in eutrophic shallow lakes. Ecological Modelling, 104(2-3): 261-287.

Bak, P., 1996. How Nature Works: Algorithms, Calibrations, Predictions. Cambridge University Press, Cambridge.

Balzter, H., Braun, P.W. and Kohler, W., 1998. Cellular automata models for vegetation dynamics. Ecological Modelling, 107(2-3): 113-125.

Baptist, M.J., 2005. Modelling floodplain biogeomorphology, Phd thesis, Delft University of Technology, Delft, 193 pp.

Batty, M., Xie, Y. and Sun, Z., 1999. Modeling urban dynamics through GIS-based cellular automata. Computers, Environment and Urban Systems, 23: 205-233.

Berger, T., 2001. Agent-based spatial models applied to agriculture: a simulation tool for technology diffusion, resource use changes and policy analysis. Agric. Econ. , 25(2/3): 245-260.

Berger, U., Piou, C., Schiffers, K. and Grimm, V., 2008. Competition among plants: concepts, individual-based modelling approaches, and a proposal for a future research strategy. Perspectives in Plant Ecology, Evolution and Systematics, 9(3-4): 121-135.

Bezdec, J.C., 1981. Pattern recognition with fuzzy objective function algorithms, New York.

Bigelow, J.H., Bolten, J.G. and de Haven, J.C., 1977. Protecting an estuary from floods - A policy analysis of the Oosterschelde, vol IV: Assessment of Algae Blooms, a potential ecological disturbance. R-212/4-NETH, 4. Rand Corporation - Rijkswaterstaat, Delft.

Blauw, A., Los, H., Bokhorst, M. and Erftemeijer, P., 2009. GEM: a generic ecological model for estuaries and coastal waters. Hydrobiologia, 618(1): 175-198.

Blauw, A.N. et al., 2006. The use of fuzzy logic for data analysis and modelling of European harmful algal blooms: results of the HABES project. African Journal of Marine Science, 28(2): 365-369.

Blonda, P.N., et al., 1991. An experiment for the interpretation of multitemporal remotely sensed images based on a fuzzy logic approach. International Journal of Remote Sensing, 12(3): 463-476.

Bobbin, J. and Recknagel, F., 2001. Inducing explanatory rules for the prediction of algal blooms by genetic algorithms. Environment International 27(2-3): 237-242.

Bonabeau, E., 2002. Agent-based modeling: Methods and techniques for simulating human systems. Proceedings of the National Academy of Sciences of the United States of America, 99(Suppl 3): 7280-7287.

Bricaud, A., Bosc, E. and Antoine, D., 2002. Algal biomass and sea surface temperature in the mediterranean Basin: Intercomparison of data from various satellite sensors, and implications for primary production estimates. Remote Sensing of Environment, 81(2-3): 163-178.

Brönmark, C. and Hansson, L.-A., 2005. The biology of lakes and ponds. Biology of habitats, xiv. Oxford University Press, Oxford, 285 pp.

Carpenter, S., Brock, W. and Hanson, P., 1999. Ecological and social dynamics in simple models of ecosystem management. Conservation Ecology, 3(2:4).

Carr, G.M., Duthie, H.C. and Taylor, W.D., 1997. Models of aquatic plant productivity: a review of the factors that influence growth. Aquatic Botany, 59(3-4): 195-215.

Chen, Q., 2004. Cellular automata and artificial intelligence in ecohydraulics modelling, UNESCO-IHE, TU Delft, Delft, 152 pp.

Chen, Q., Morales-Chaves, Y., Li, H. and Mynett, A.E., 2006. Hydroinformatics techniques in eco-environmental modelling and management. Journal of Hydroinformatics, 08(4): 297-316.

Chen, Q., Mynett, A.E. and Minns, A.W., 2002. Application of cellular automata to modelling competitive growths of two underwater species Chara aspera and Potamogeton pectinatus in Lake Veluwe. Ecological Modelling, 147(3): 253-265.

Chen, Q.H., Zeng, Z., Zhang, S. and Xu, K., 1993. Report on the Red Tide occurred in Xiamen Harbour in 1987. In: Third Institute of Oceanography State Oceanic Administration (Editor), Collected papers on red tidesurvey and study in Xiamen Harbour. China Ocean Press, Beijing.

Cheng, K., et al, 2000. An anisotropic spatial modeling approach for Remote Sensing image rectification. Elsevier, Remote Sensing and Environment, 73: 46-54.

Chiarello, E. and Barrat-Segretain, M.H., 1997. Recolonization of cleared patches by macrophytes: modelling with point processes and random mosaics. Ecological modelling, 96: 61-73.

Clark, L.R., Geier, P.W., Hughes, R.D. and Morris, R.F., 1967. The ecology of insect populations in theory and practice. Methuen, London.

Cole, V. and Albrecht, J., 1999. Exploring Geographic Parameter Space With a GIS Im-plementation of Cellular Automata, Proceedings of SIRC'99, The 11th Annual Colloquium of the Spatial Information Research Centre.

Connor, E.F. and McCoy, E.D., 1979. The statistics and biology of the species-area relationship. The American Naturalist, 113(6): 791.

Connors, K.A., 1998. Chemical kinetics: the study of reaction rates in solution, 496 pp.

Coops, H. and Doef, R.W., 1996. Submerged vegetation development in two shallow, eutrophic lakes. Hydrobiologia, 340(1): 115-120.

Czaran, T. and Bartha, S., 1992. Spatiotemporal dynamic models of plant populations and communities. Trends in Ecology & Evolution, 7(2): 38-42.

Czogala, E. and Leski, J., 2000. Fuzzy and Neuro-Fuzzy intelligence systems Studies in fuzziness and soft computing. Physica-Verlag, Heidelberg, Germany.

Dale, M.R.T., 2004. Spatial pattern analysis in plant ecology. Cambridge University Press, 338 pp.

Davidsson, P., Holmgren, J., Kyhlbäck, H., Mengistu, D. and Persson, M., 2007. Applications of agent based simulation, Multi-Agent-Based Simulation VII, pp. 15-27.

DeAngelis, D.L. and Mooij, W.M., 2005. Individual-based modelling of ecological and evolutionary processes. Annual Review of Ecology, Evolution, and Systematics, 36(1): 147-168.

Di Toro, D.M., O'Connor, D.J. and Thomann, R.V., 1971. A dynamic model of the phytoplankton population in the Sacramento-San Joaquin Delta. In: R.F. Gould (Editor), Nonequilibrium Systems in Natural Water Chemistry. Advances in Chemistry series. American Chemical Society, Wash. DC, pp. 131-180.

Edelvang, K. et al., 2005. Numerical modelling of phytoplankton biomass in coastal waters. Journal of Marine Systems, 57(1-2): 13-29.

Ferber, J., 1999. Multi-Agent Systems: An introduction to distributed Artificial Intelligence. Addison-Wesley professional.

Field, C., Behrenfeld, M., Randerson, J. and Falkowski, P., 1998. Primary production of the biosphere: integrating terrestrial and oceanic components. Science, 281(5374): 237-240.

Franks, P., 1997. Spatial patterns in dense algal blooms. Limnology and Oceanography Journal, 42(5): 1297-1305.

Freckleton, R.P. and Watkinson, A.R., 2002. Large-scale spatial dynamics of plants: metapopulations, regional ensembles and patchy populations. Journal of Ecology, 90(3): 419-434.

Gardner, M., 1970. Mathematical games: the fantastic combinations of John Conway's new solitaire game "life". Scientific American(223): 120-123.

Gardner, R.H. and Engelhardt, K.A.M., 2008. Spatial processes that maintain biodiversity in plant communities. Perspectives in Plant Ecology, Evolution and Systematics, 9(3-4): 211-228.

Gerritsen, H., Vos, R.J., van der Kaaij, T., Lane, A. and Boon, J.G., 2000. Suspended sediment modelling in a shelf sea (North Sea). Coastal Engineering, 41: 317-352.

Gilbert, N. and Troitzsch, K.G., 2005. Simulation for the social scientist. Open University Press, New York, 295 pp.

Giusti, E. and Marsili-Libelli, S., 2006. An integrated model for the Orbetello lagoon ecosystem. Ecological Modelling, 196(3-4): 379-394.

Griffiths, M., 2002. The European water framework directive: an approach to integrated river basin management, European Water Management Online. European Water Management, pp. 14.

Grimm, V., 1999. Ten years of individual-based modelling in ecology: what have we learned and what could we learn in the future? Ecological Modelling, 115(2-3): 129-148.

Grimm, V. and Railsback, S.F., 2005. Individual-based modeling and ecology. Princeton Series in Theoretical and Computational Biology. Princeton University Press.

Grimm, V. et al., 2005. Pattern-Oriented Modeling of Agent-Based complex systems: lessons from ecology. Science, 310(5750): 987-991.

Haan, C.T., 1977. Statistical methods in hydrology. The Iowa State University Press, 378 pp.

Hakanson, L., 1999. On the principles and factors determining the predictive success of ecosystem models, with a focus on lake eutrophication models Ecological Modelling, 121(2): 139-160.

Hanski, I. and Simberloff, D., 1997. The metapopulation approach, its history, conceptual domain, and application to conservation. In: I. Hanski and M.E. Gilpin (Editors), Metapopulation Biology: Ecology, Genetics and Evolution. Academic Press, New York, pp. 5-26.

Hogeweg, P., 1988. Cellular automata as a paradigm for ecological modeling. Appl. Math. Comput., 27(1): 81-100.

Hogeweg, P., 2007. From population dynamics to ecoinformatics: Ecosystems as multilevel information processing systems. Ecological Informatics, 2(2): 103-111.

Holmes, E.E., Lewis, M.A., Banks, J.E. and Veit, R.R., 1994. Partial differential equations in ecology: spatial interactions and population dynamics. Ecology, 75: 17-29.

Hosper, H., 1997. Clearing lakes, PhD Thesis, Agricultural University, Wageningen, The Netherlands, 168 pp.

Hovel, K. and Regan, H., 2008. Using an individual-based model to examine the roles of habitat fragmentation and behavior on predator–prey relationships in seagrass landscapes. Landscape Ecology, 23(Suppl. 1): 75-89.

Huston, M.A., 1994. Biological diversity: the coexistence of species on changing landscapes. Cambridge University Press, Cambridge.

Ilachinski, A., 2001. Cellular automata: a discrete universe, xxxii. World Scientific, Singapore ; River Edge, NJ, 808 pp.

Jeltsch, F. and Moloney, K., 2002. Spatially-explicit vegetation models: what have we learned? In: K. Esser, U. Luettge, W. Beyschlag and F. Hellwig (Editors). Progress in Botany. Springer Verlag, Berlin, pp. 326-343.

Jeltsch, F., Moloney, K.A., Schurr, F.M., Kochy, M. and Schwager, M., 2008. The state of plant population modelling in light of environmental change. Perspectives in Plant Ecology, Evolution and Systematics, 9(3-4): 171-189.

Jørgensen, S.E., 2008. Overview of the model types available for development of ecological models. Ecological Modelling, 215(1-3): 3-9.

Jørgensen, S.E. and Bendoricchio, G., 2001. Fundamentals of ecological modelling. Developments in environmental modelling, xii. Elsevier, Amsterdam ; London, 530 pp.

Joshua, M.E. and Robert, A., 1996. Growing artificial societies: social science from the bottom up. The Brookings Institution, 208 pp.

Kantrud, H.A., 1990. Sago pondweed (Potamogeton pectinatus L.): A literature review. Fish and Wildlife Resource Publication 176, Jamestown, ND: Northern Prairie Wildlife Research Center online http://www.npwrc.usgs.gov/resource/plants/pondweed/index.htm. (Version 16 JUL 97).

Koikkalainen, P. and Horppu, I., 2007. Handling missing data with the tree-structured Self-Organizing Map, Proceedings of International Joint Conference on Neural Networks, Orlando, Florida, USA, August 12-17, 2007, pp. 1-6.

Lee, J.H.W., Huang, Y., Dickman, M. and Jayawardena, A.W., 2003. Neural network modelling of coastal algal blooms. Ecological Modelling, 159(2-3): 179-201.

Lee, J.H.W. and Qu, B., 2004. Hydrodynamic tracking of the massive spring 1998 red tide in Hong Kong. Journal of Environmental Engineering, 130(5): 535-550.

Legendre, P. and Fortin, M.J., 1989. Spatial pattern and ecological analysis. Plant Ecology, 80(2): 107-138.

Legendre, P. and Legendre, L., 1998. Numerical Ecology. Elsevier, Amsterdam.

Lesser, V.R., 1999. Cooperative multiagent systems: a personal view of the state of the art. IEEE Transactions on Knowledge and Data Engineering, 11(1): 133-142.

Levenberg, K., 1944. A method for the solution of certain problems in least squares. Quart. Appl. Math, 2: 164-168.

Levin, S., 1992. The problem of pattern and scale in ecology. ecology, 73(6): 1943-1967.

Levin, S.A., 1976. Population dynamic models in heterogeneous environments. Annual Review of Ecology and Systematics, 7(1): 287-310.

Levins, R., 1966. The strategy of model building in population biology. American Scientist, 54: 421-431.

Levins, R., 1970. Extinction. In: M. Gesternhaber (Editor), Some Mathematical Problems in Biology. American Mathematical Society, Providence, Rhode Island. , pp. 77-107.

Li, H., 2005. Harmful Algal Bloom Prediction, a case study for Western Xiamen Bay, China. MSc Thesis, UNESCO-IHE, Institute for water education, Delft, The Netherlands, 78 pp.

Li, H., Arias, M., Blauw, A., Peters, S. and Mynett, A.E., 2009a. Enhancing Delft3D-BLOOM/GEM for algae spatial pattern analysis: model improvement, Proceedings of the International Conference on "Science and Information Technologies for Sustainable Management of Aquatic Ecosystems", the joint meeting of the 7th International Symposium on Ecohydraulics and the 8th International Conference on Hydroinformatics, Concepcion, Chile

Li, H., Mynett, A.E., Arias, M., Blauw, A. and Peters, S., 2008a. A Pilot study for an enhanced algal spatial pattern prediction using Remote Sensing images. In: C. Zhang and H. Tang (Editors), Proc.16th

IAHR-APD Congress. Tsinghua University Press, Nanjing, China, pp. 738-743.

Li, H., Mynett, A.E. and Chen, Q., 2006a. Modelling of algal population dynamics using cellular automata and fuzzy rules, Proc. 7th Int. Conf. on Hydroinformatics. Research Publishing, Nice, France, pp. 1040-1047.

Li, H., Mynett, A.E. and Corzo, G., 2007a. Model-based training of Artificial Neural Networks and Cellular Automata for rapid prediction of potential algae blooms, 6th International Symposium on Ecohydraulics, Christchurch, New Zealand.

Li, H., Mynett, A.E., Huang, B.Q. and Hong, H.S., 2007b. Main factor selection in Harmful Algal Bloom Prediction with a case study for Western Xiamen Bay. In: L. Ren (Editor), Methodology in Hydrology. IAHS publication, 0144-7815. IAHS, Wallingford, pp. 345-351.

Li, H., Mynett, A.E., Huang, B.Q. and Qiuwen, C., 2006b. Harmful Algal Bloom Prediction using Data-Driven modelling: a case for Western Xiamen Bay of China, The Asia Oceania Geosciences Society's 3rd Annual Meeting (AOGS 2006). AOGS, Singapore.

Li, H., Mynett, A.E. and Penning, E., 2008b. Aquatic plant dynamics modelling using photographic based cellular automata, 6th International Conference on Ecological Informatics, Cancun, Mexico.

Li, H., Mynett, A.E. and Penning, E., 2009b. Photography-based cellular automata in aquatic plant dynamics modelling. Submitted to Journal of Ecological Informatics

Li, H., Mynett, A.E., Qi, H. and Penning, E., 2009c. Multi-Agent Systems in modelling aquatic population dynamics in Lake Veluwe, Netherlands. Submitted to Journal of Ecological informatics

Li, W., Qiang, S. and Chen, K., 2002. Photosynthetic rate of Potamogeton pectinatus L. and influencing factors (in Chinese). Journal of Lake sciences, 4.

Los, F.J., Villars, M.T. and Van der Tol, M.W.M., 2008. A 3-dimensional primary production model (BLOOM/GEM) and its applications to the (southern) North Sea (coupled physical-chemical-ecological model). Journal of Marine Systems, 74(1-2): 259-294.

Los, F.J. and Wijsman, J.W.M., 2007. Application of a validated primary production model (BLOOM) as a screening tool for marine, coastal and transitional waters. Journal of Marine Systems, 64(1-4): 201-215.

Los, H., 2009. Eco-hydrodynamic modelling of primary production in coastal waters and lakes using BLOOM, Wageningen University, 276 pp.

Lotka, A.J., 1925. Elements of physical biology. Williams & Wilkins Company, Baltimore.

Madsen, J.D., Chambers, P.A., James, W.F., Koch, E.W. and Westlake, D.F., 2001. The interaction between water movement, sediment dynamics and submersed macrophytes. Hydrobiologia, 444(1): 71-84.

Maier, H.R. and Dandy, G.C.V.p., 1997. Modelling cyanobacteria (blue-green algae) in the River Murray using artificial neural networks. Mathematics and Computers in Simulation, 43: 377-386.

Marquardt, D., 1963. An algorithm for least-squares estimation of nonlinear parameters SIAM. SIAM Journal on Applied Mathematics, 11: 431-441.

May, R.M., 1993. The effects of spatial scale on ecological questions and answers. In: P.J. Edwards, R.M. May and N.R. Webb (Editors), Large-Scale Ecology and Conservation Biology. Blackwell, Oxford, pp. 1-18.

May, R.M. and Oster, G.F., 1976. Bifurcations and dynamic complexity in simple ecological models. The American Naturalist, 110(974): 573.

Milbradt, P. and Schonert, T., 2008. A holistic approach and object-oriented framework for eco-hydraulic simulation in coastal engineering. Journal of Hydroinformatics, 10(3): 201-214.

Minns, A.W., Mynett, A.E., Chen, Q. and van den Boogaard, H.F.P., 2000. A cellular automata approach to ecological modelling. In: A.J. Odgaard (Editor), Proceedings of the 4th International Conference on Hydroinformatics. Cedar Rapids, Iowa, USA.

Moll, A. and Radach, G., 2003. Review of three-dimensional ecological modelling related to the North Sea shelf system: Part 1: models and their results. Progress In Oceanography, 57(2): 175-217.

Morales, Y., Weber, L.J., Mynett, A.E. and Newton, T.J., 2006. Mussel Dynamics Model: a hydroinformatics tool for analyzing the effects of different stressors on the dynamics of freshwater mussel communities. Ecological Modelling, 197(3-4): 448-460.

Murray, S., 1993. Neural networks for statistical modelling. Van Nostrand Reinhold Press, New York.

Mynett, A. and Chen, Q., 2004. Cellular automata in ecological and ecohydraulics modelling. In: P.M.A. Sloot, B. Chopard and A.G. Hoekstra (Editors), 6th International Conference on Cellular Automata for Research and Industry. Springer, Amsterdam, the Netherlands, pp. 502-512.

Mynett, A.E., 1999. Art of modelling - water systems in their natural environment;inaugural address, delivered on the occasion of the public acceptance of the Chair in Environmental Hydroinformatics, International Institute for Infrastructural, Hydraulic and Environmental Engineering (IHE), Delft, The Netherlands,.

Mynett, A.E., 2002. Environmental Hydroinformatics: the way ahead. In: Falconer (Editor), Proceedings of the Fifth International Conference on Hydroinformatics. IWA, Cardiff, UK, pp. 31-36.

Mynett, A.E., 2004. Hydroinformatics tools for ecohydraulics modelling, opening keynote. In: P. Liong and V. Babovic (Editors), Proc 6th Int. Conf. on Hydroinformatics World Scientific Publishing, Singapore, pp. 3-12.

Mynett, A.E., Lin, Y. and Chen, Q., 2009. Unstructured cellular automata for enhanced population dynamics modelling, Proceedings of the International Conference on "Science and Information Technologies for Sustainable Management of Aquatic Ecosystems", the joint meeting of the 7th International Symposium on Ecohydraulics and the 8th International Conference on Hydroinformatics, Concepcion, Chile.

Mynett, A.E. and Morales, Y., 2006. Individual based modelling in ecosystem dynamics, Proc. 7th Int. Conf. on Hydroinformatics. Research Publishing, Nice, pp. 1399-1406.

National Science and Technology Council Committee on Environment and Natural Resources, 2000. National assessment of harmful algal blooms in US waters, National Science and Technology Council Committee on Environment and Natural Resources.

Paarlberg, A.J., Knaapena, M.A.F., de Vries, M.B., Hulschera, S.J.M.H. and Wang, Z.B., 2005. Biological influences on morphology and bed composition of an intertidal flat. Estuarine, Coastal and Shelf Science, 64(4): 577-590.

Pacala, S.W., 1997. Dynamics of plant communities. In: M.J. Crawley (Editor), Plant Ecology, 2nd edition. Blackwell Science, Ltd., Cambridge, pp. 532-555.

Packard, N.H. and Wolfram, S., 1985. Two-dimensional cellular automata. Journal of Statistical Physics, 38: 901-946.

Park, G.S. and Park, S.Y.V.p., 2000. Long-term trends and temporal heterogeneity of water quality in tidally mixed estuarine waters. Marine Pollut. Bull., 40: 1201-1209.

Park, S. and Sugumaran, V., 2005. Designing multi-agent systems: a framework and application. Expert Syst. Appl., 28(2): 259-271

Parker, D.C., Manson, S.M., Janssen, M.A., Hoffmann, M.J. and Deadman, P., 2003. Multi-Agent Systems for the simulation of land-use and land-cover change: a review. Annals of the Association of American Geographers, 93(2): 314-337.

Parrott, L., 2005. Quantifying the complexity of simulated spatiotemporal population dynamics. Ecological Complexity, 2(2): 175-184.

Parrott, L. and Kok, R., 2006. Use of an object-based model to represent complex features of ecosystems, Unifying Themes in Complex Systems, pp. 169-179.

Pasterkamp, R. and van der Woerd, H.J., 2008. HYDROPT: a fast and flexible method to retrieve chlorophyll-a from multi-spectral satellite observation of optical-complex coastal waters. Remote sensing of environment, 112(4): 1795-1807.

Peperzak, L., Colijn, F., Gieskes, W.W.C. and Peeters, J.C.H., 1998. Development of the diatom *phaeocystis* spring bloom in the Dutch coastal zone of the North Sea: the sili-con depletion versus the daily irradiance threshold hypothesis. . Journal of Plankton Research, 20: 21.

Pereira, A., Duarte, P. and Norro, A., 2006. Different modelling tools of aquatic ecosystems: A proposal for a unified approach. Ecological Informatics, 1(4): 407-421.

Peterson, D.L. and Parker, V.T. (Editors), 1998. Ecological scale: theory and applications. Complexity in Ecological Systems. Columbia University Press New York.

Postma, L., 2003. Water quality of surface waters, WL|Delft Hydraulics, Delft.

Postma, L., 2007. Modelling of water quality and ecology is dealing with problems of scale. La Houille Blanche, 5: 37-42.

Price, R.K., 2006. The growth and significance of hydroinformatics, River Basin Modelling for Flood Risk Mitigation. Taylor and Francis Group plc, London, UK, pp. 93-109

Qi, Y.Z., 2003. Coastal Red tide in China (in Chinese). China Science Press, Beijing, 348 pp.

Radach, G. and Moll, A., 2006. Review of three-dimensional ecological modelling related to the North Sea shelf system. Part 2: Model validation and data needs. Oceanography and Marine Biology 44: 1-60.

Railsback, S.F., Lytinen, S.L. and Jackson, S.K., 2006. Agent-based Simulation Platforms: Review and Development Recommendations. Simulation, 82(9): 609-623.

Ratz, C., Gillet, F., Muller, J.-P. and Stoffel, K., 2007. Simulation modelling of ecological hierarchies in constructive dynamical systems. Ecological Complexity, 4(1-2): 13-25.

Recknagel, F., Bobbin, J., Whigham, P. and Wilson, H., 2002. Comparative application of artificial neural networks and genetic algorithms for multivariate time-series modelling of algal blooms in freshwater lakes. Journal of Hydroinformatics, 4(2): 125-134.

Recknagel, F.e., 2002. Ecological Informatics: understanding ecology by biologically-inspired computation. Springer Verlag, Berlin, Heidelberg.

Resnick, M., 1994. Turtles, termites, and traffic jams: explorations in massively parallel microworlds. Complex adaptive systems, xviii. MIT Press, Cambridge, Mass, London, 163. pp.

Reynolds, C., W., 1987. Flocks, herds and schools: a distributed behavioral model. Computer Graphics, 21(4)(Proceedings of the 14th annual conference on Computer graphics and interactive techniques): 25-34.

Robson, B.J. and Hamilton , D.P., 2004. Three-dimensional modelling of a Microcystis bloom event in the Swan River estuary, Western Australia. Special Issue of Ecological Modelling, 174: 203-222.

Romero, J.R., Antenucci, J.P. and Imberger, J., 2004. One- and three-dimensional biogeochemical simulations of two differing reservoirs. Ecological Modelling, 174(1-2): 143-160.

Rothman, D.H. and Zaleski, S., 1997. Lattice Gas Cellular Automata: simple models of complex hydrodynamics. Collection Alea-Saclay: Monographs and Texts in Statistical Physics. Cambridge University Press, 320 pp.

Sacau-Cuadrado, M., Conde-P, P. and Otero-Tranchero, P., 2003. Forecast of red tides off the Galiciabj coast. Acta Astronautica 53: 5.

Sandwell, D.T., 1987. Biharmonic spline interpolation of GEOS-3 and SEASAT altimeter data. Geophysical Research Letters, 14(2): 139-142.

Schaefer, J.T., 1990. The critical success index as an indicator of warning skill. Weather and Forecasting, 5(4): 570-575.

Scheffer, M., Bakema, A. and Wortelboer, F., 1993. MEGAPLANT: a simulation model of the dynamics of submerged plants. . Aquatic Botany 45(4): 341-356.

Schneider, D., 1994. Quantitative ecology: spatial and temporal scaling Academic Press, 395 pp.

Schönfisch, B. and de Roos, A., 1999. Synchronous and asynchronous updating in cellular automata. Biosystems, 51(3): 123-143.

Seppelt, R., 2002. Avenues of spatially explicit population dynamics modeling - a par excellence example for mathematical heterogeneity in ecological models?, Integrated Assessment and Decision Support. IDSI, Switzerland, pp. 269-274

Sharov, A.A., 1992. The life-system approach: a system paradigm in population ecology. Oikos, 63(3): 485-494.

Shnerb, N.M., Louzoun, Y., Bettelheim, E. and Solomon, S., 2000. The importance of being discrete: life always wins on the surface. Proceedings of the National Academy of Sciences, 97(19): 10322-10324.

Silvertown, J., Holtier, S., Johnson, J. and P., D., 1992. Cellular automaton models of interspecific competition for space: effect of pattern on process. Journal of Ecology, 80: 527–534.

Simon, H.A., 1996. The Sciences of the Artificial MIT Press, 247 pp.

Solomatine, D.P., Abrahart, R. and L., S., 2008. Data-driven modelling: concept, approaches, experiences. In: Abrahart, See and Solomatine (Editors), Practical Hydroinformatics: Computational Intelligence and Technological Developments in Water Applications. Springer-Verlag, Berlin.

Stanski, H., R.,, Wilson, L.J. and Burrows, W.R., 1989. Survey of common verification methods in meteorology, WMO World Weather Watch Technical Report No. 8, WMO/TD No. 358, Atmospheric Environment Service, Forecast Research Division, Geneva.

Stelling, G.S., 1984. On the construction of computational methods for shallow water flow problems. In: R. communications (Editor). Rijkswaterstaat, The Hague.

Suzudo, T., 2004. Searching for pattern-forming asynchronous Cellular Automata-an evolutionary approach, Proceedings in the 6th international conference on Cellular Automata for research and industry, ACRI 2004. Springer-Verlag Berlin Heidelberg, pp. 151-160.

Sycara, K.P., 1998. The many faces of agents. AI Magazine, 19(2): 11-12

Temmerman, S. et al., 2007. Vegetation causes channel erosion in a tidal landscape. Geology, 35(7): 631-634.

Tilman, D. and Kareiva, P. (Editors), 1997. Spatial ecology : the role of space in population dynamics and interspecific interactions Monographs in population biology. Princeton University Press, New Jersey.

Toffoli, T. and Margolus, N., 1987. Cellular automata machines : a new environment for modeling, ix. MIT Press, Cambridge, Mass, London, 259 pp.

Trancoso, A.R. et al., 2005. Modelling macroalgae using a 3D hydrodynamic-ecological model in a shallow, temperate estuary. Ecological Modelling, 187(2-3): 232-246.

Troitzsch, K.G., 1997. Social science simulation-origins, prospects, purposes. In: R. Conte, R. Hegselmann and P. Terno (Editors), Simulating Social Phenomena. Springer-Verlag, Berlin, pp. 41-54.

Trunfio, G.A., 2004. Predicting wildfire spreading through a hexagonal cellular auto-mata model, Proceedings in the 6th international conference on Cellular Automata for research and industry, ACRI 2004. Springer-Verlag Berlin Heidelberg, pp. 385-394.

Uittenbogaard, R., 2003. modelling turbulence in vegetated aquatic flows, International workshop on Riparian Forest vegetated channels: hydraulic, morphological and ecological aspects, Trento, Italy.

United Nations, 2008. The Millennium Development Goals report 2008, United Nations, New York.

Urbanski, J.A., 1999. The use of fuzzy sets in the evaluation of the environment of coastal waters. int. j. Geographical information science, 13(7): 723- 730.

Van den Berg, M.S., 1999. Charophyte colonization in shallow lakes: process, ecological effects and implications for lake management, Vrije Universiteit Amsterdam, Amsterdam, the Netherlands, 138 pp.

Van den Berg, M.S., Coops, H. and Simons, J., 2001. Propagule bank buildup of Chara aspera and its significance for colonization of a shallow lake. Hydrobiologia, 462(1): 9-17.

Van den Berg, M.S., Scheffer, M., Coops, H. and Simons, J., 1998. The role of characean algae in the management of eutrophic shallow lakes. Journal of Phycology, 34(5): 750-756.

Van den Berg, M.S., Scheffer, M., Van Nes, E.H. and Coops, H., 1999. Dynamics and stability of Chara sp. and P. pectinatus in a shallow lake changing in eutrophication level. Hydrobiologia, 408/409: 335-342.

Van der Woerd, H.J. et al., 2005. Integrated spatial and spectral characterisation of harmful algal blooms in Dutch coastal waters (ISCHA): demonstration of a HAB service in the Zeeuwse Voordelta, Institute for environmental studies, Vrije Universiteit, Amsterdam.

Van der Woerd, H.J. and Pasterkamp, R., 2008. HYDROPT: a fast and flexible method to retrieve chlorophyll-a from multispectral satellite observations of optically complex coastal waters. Remote Sensing of Environment, 112(4): 1795-1807.

Van Nes, E.H., Scheffer, M., van den Berg, M.S. and Coops, H., 2003. Charisma: a spatial explicit simulation model of submerged macrophytes. Ecological Modelling, 159(2-3): 103-116.

Velez, C.A. and Mynett, A.E., 2006. Water quality and ecosystem modelling - a case study for Sonso Lagoon, Colombia, Proc. 7th Int. Conf. on Hydroinformatics. Research Publishing, Nice, France, pp. 1803-1810.

Verhulst, P.F., 1838. Notice sur la loi que la population poursuit dans son accroissement. Correspondance mathématique et physique, 10: 113-121.

Vlassis, N., 2007. A concise introduction to Multiagent Systems and Distributed Artificial Intelligence. Synthesis Lectures in Artificial Intelligence and Machine Learning. Morgan & Claypool Publishers, 71 pp.

Vollenweider, R.A., 1975. Input-output models, with special reference to the phosphorus loading concept in limnology. Schweiz. Zeitschr. Hydrol., 37: 53-84.

Volterra, V., 1931. Variations and fluctuations of the number of individuals in animal species living together (Translated from 1928). In: R.N. Chapman (Editor), Animal Ecology.

Von Altrock, C., 1995. Fuzzy logic and neurofuzzy applications explained. Prentice-Hall, Inc., 350 pp.

Von Neumann, J., 1949. Theory of self-reproducing automata. University of IlliNoise Press, Urbana and London, 388 pp.

Wagner, D.F., 1997. Cellular automata and geographic information systems. Environment and Planning B: Planning and Design, 24: 219-234.

Wang, W., 2006. Stochasticity, nonlinearity and forecasting of streamflow processes. IOS Press, the Netherlands, 210 pp.

Watt, A.S., 1947. Pattern and process in the plant community. Journal of Ecology, 35: 1-22.

Weber, L.J., Goodwin, R.A., Li, S., Nestler, J.M. and Anderson, J.J., 2006. Application of an Eulerian–Lagrangian–Agent method (ELAM) to rank alternative designs of a juvenile fish passage facility. Journal of Hydroinformatics, 08(4): 271-295.

Weidenhamer, J.D., 1996. Distinguishing resource competition and chemical interference: overcoming the methodological impasse. Agron J, 88(6): 866-875.

Weiss, S.M. and Indurkhya, N., 1998. Predictive data mining, a practical guide. Morgan Kaufmann Publishers, 228 pp.

Weisstein, E., 2008. Wolfram MathWorld, Online Mathematics Encyclopedia. Wolfram Research.

Werner, M., Blazkova, S. and Petr, J., 2005. Spatially distributed observations in constraining inundation modelling uncertainties. Hydrological Processes, 19(16): 3081-3096.

Wetzel, R.G., 2001. Limnology: lake and river ecosystems Academic Press, San Diego, USA, 850 pp.

Wiegand, T., Jeltsch, F., Hanski, I. and Grimm, V., 2003. Using pattern-oriented modeling for revealing hidden information: a key for reconciling ecological theory and application. Oikos, 100(2): 209-222.

Witten, I.H. and Frank, E., 2005 Data mining: practical machine learning tools and techniques. Morgan Kaufmann Series in Data Management Systems, 525 pp.

WL | Delft Hydraulics, 2006a. Delft3D-FLOW users manual, WL | Delft Hydraulics, Delft, the Netherlands.

WL | Delft Hydraulics, 2006b. Open process library user manual, WL | Delft Hydraulics, Delft, the Netherlands.

WL | Delft Hydraulics, 2007. Delft3D-WAQ user manual: Versatile water quality modelling in 1D, 2D or 3D systems including physical, (bio)chemical and biological processes, WL | Delft Hydraulics, Delft.

Wolfram, S., 1986. Theory and applications of cellular automata : including selected papers 1983-1986. World Scientific.

Wolfram, S., 2002. A new kind of science. Wolfram Media London, 1197 pp.

Wootton, J.T., 2001. Local interactions predict large-scale pattern in empirically derived cellular automata. Nature, 413(6858): 841-844.

Yokozawa, M., Kubota, Y. and Hara, T., 1998. Effects of competition mode on spatial pattern dynamics in plant communities. Ecological Modelling, 106(1): 1-16.

Zadeh, L.A., 1965. Fuzzy sets. Information and Control, 8(3): 338-353.

Zhang, S.J., 1993. Ecological features of phytoplankton from red tide occurring area in Xiamen Harbour. In: Third Institute of Oceanography State Oceanic Administration (Editor), Collected papers on red tide survey and study in Xiamen Harbour (in Chinese). China Ocean Press, Beijing, pp. 29-36.

# List of figures

# List of tables

# List of symbols and abbreviations

| Symbol | Unit | Description |
|---|---|---|
| $NH_4^+$ | | ammonia |
| $NO_2^-$ | | nitrite |
| $NO_3^-$ | | nitrate |
| $D_{x,y}$ | (m²/s) | dispersion coefficient |
| $C$ | | substance concentration |
| $u, v$ | (m/s) | components of the velocity vector |
| $S$ | | sources and sinks of mass due to loads and boundaries |
| $P$ | | sources and sinks of mass due to processes |
| $\nabla$ | | nabla operator |
| $r$ | | coefficient of correlation $r$ |
| $x , y$ | | two variables $x$ and $y$, |
| $\bar{x}\ \bar{y}$ | | the mean of variable $x$, the mean of variable $y$. |
| $f$ | | membership function in FL model |
| $dt, \Delta t, DELT$ | | model time step |
| $I_{1t}, I_{2t}$ | | inputs for FL model |
| $L$ | | a lattice in CA |
| $Q$ | | a state space |
| $N$ | | neighbourhood scheme in CA |
| $f$ | | local transition function in CA |
| $S$ | | cell states in CA |
| $P_{1t}, P_{2t}, P_{3t}\ldots$ | | other factors which are included as model inputs |
| $X_{i+1, j}, X_{i-1,j}, X_{i,j+1}, X_{i,j-1}$ | | values of the nearest neighbouring cells in CA |
| plant_age | week | how many weeks the plant already grew |
| maximum_lifespan | week | the maximum possible life span when the plant is in an optimal environment. |
| maximum_height | m | the maximum height specific aquatic plant can grow |
| height(t) | m | plant height at time t |
| $D_t$ | | aquatic plant density at certain location in time t |
| $D_g, D_{sg}, D_e, D_i, D_m, D_{sd}$ | | contributions from different processes to total aquatic plant density |
| $\Delta$ | | flux in one time step based on density function |
| $R$ | | small random variation due to stochasticity in the system |
| $P$ | | random probability involved in the spatial density variation |
| $K$ | | carrying capacity of certain area |
| $r, r\_cs, r\_pp$ | | growth rate, growth rate for Pp and Cs |
| $R\_birds$ | | randomness in seed dispersal due to birds |
| StepToLive | week | how many weeks left for a plant to be alive |
| $A$ | | total number of cells correctly predicted |

| | | |
|---|---|---|
| B | | total number of over-estimated cells |
| C | | total number of under-estimated cells |
| Np_max | stem/m$^2$ | carrying capacity |
| Np | stem/m$^2$ | stem density of aquatic plant |

| Abbreviation | Description |
|---|---|
| ABM | Agent-Based Model |
| AI | Artificial Intelligence |
| ANNs | Artificial Neural Networks |
| BP | Back Propagation algorithm |
| CA | Cellular Automata |
| Chl-a | Chlorophyll a |
| Cs | Chara aspera |
| DAI | distributed Artificial Intelligence |
| DDM | Data-Driven Model |
| DIP | Dissolved Inorganic Phosphorus |
| DLL | dynamic link library |
| DO | Dissolved Oxygen |
| DOP | Dissolved Organic Phosphorus |
| FCM | Fuzzy C-Means clustering |
| FL | Fuzzy Logic |
| GEM | Generic Ecological Model |
| GIS | Geographic Information System |
| HAB | Harmful Algal Bloom |
| IBE | Individual Based Ecology |
| IBM | Individual Based Model |
| LV | Lotka-Voterra |
| MAE | Mean Absolute Error |
| MAS | Multi-Agent Systems |
| MERIS | MEdium Resolution Imaging Spectrometer instrument |
| MLP | Multi-Layer Perceptron neural network |
| ODE | Ordinary Differential Equations |
| OPL | Open Process Library in DELWAQ |
| PB | Physically-Based model |
| PCA | Principal Component Analysis |
| PDE | Partial Differential Equations |
| PoD | Probability of Detection |
| Pp | Potamogeton pectinatus |
| R$^2$ | linear correlation coefficient |
| RMSE | Root Mean Squared Error |
| RS | Remote Sensing |
| SLCA | Self-Learning Cellular Automata |
| TIN | Total Inorganic Nitrogen |
| TIP | Total Inorganic Phosphorous |
| TSM | Total Suspended Matters |
| WTD | Water Temperature weekly Difference |

# Appendix

## I Comparison of chlorophyll-a results from RS images, original model and enhanced model

### Introduction

This appendix is an extension of chapter 5 and includes a more detailed comparison of Chlorophyll-a (Chl-a) maps retrieved from MERIS images, the computational results from the original BLOOM/GEM model, and from the enhanced model using TSM retrieved from MERIS data as model inputs.

### Chl-a maps

*Retrieved from RS*          *Original model*          *Enhanced model*

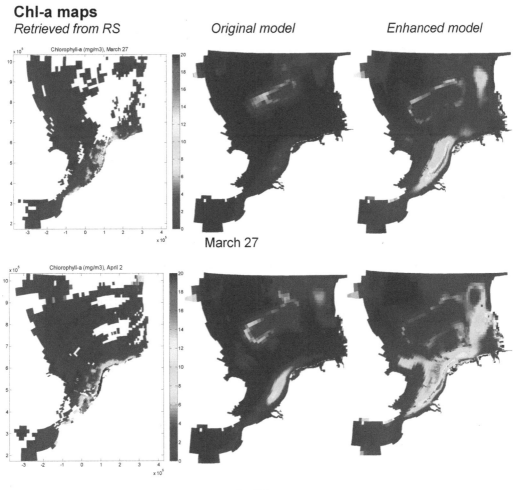

March 27

April 2

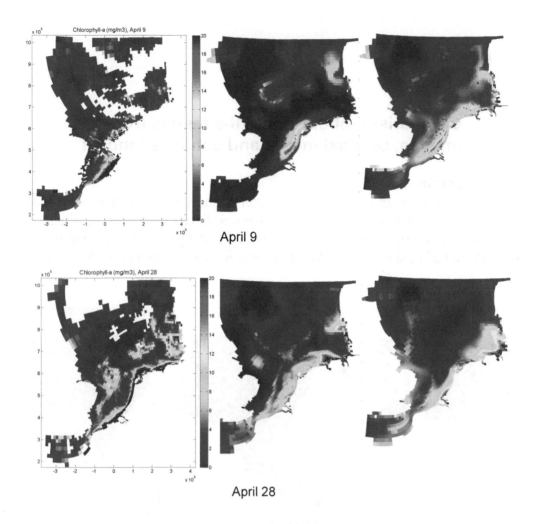

April 9

April 28

The results clearly indicate that a better representation of spatial patterns for Chl-a can be obtained by enhancing the original BLOOM/GEM model with spatially distributed input data retrieved from MERIS satellite measurements of Total Suspended Matter (TSM).

## II    Cellular Automata modelling for water lily growth

### Introduction

The model introduced in Chapter 6 combines weekly high resolution photos and traditional Cellular Automata for simulating water lily growth in a small and relatively still (no water flow) pond. The pre-processing of high resolution photos was carried out in Coral Photoshop and ArcGIS, and the model was coded in Matlab with online animation of modelling results.

### Rules and pseudo codes

The states considered in this model are (i) water and (ii) water lily. The initial water lily distribution map was derived from one photo taken in the middle of April (wk 18), 2007. The main environmental factors considered here are weekly averaged temperature, weekly accumulated sunlight duration and temperature differences between each week. Water lilies grow, extend and decay based on their current cell state and the neighbouring cell states under the constraints of environmental factors such as sunlight duration and temperature. The rules and functions for cell state updates are formulated below in pseudo-code, where:

*temp* is weekly averaged temperature,

*sd* is weekly accumulated sunshine duration,

*temp_diff* is the temperature difference between present and previous week,

*sum_nei* is the number of neighbouring cells which contain water lily.

*Young* water lily here means younger than 2/3 of maximum lifespan.

### If cell is water:

*% optimal conditions for new lily:*

*e.g. temp > 16, sd > 40 and sum_nei>=1 and the neighbouring lily is still young*

*=> cell grows new lily*

*% other situations for newly growth:*

*when temp decreases, or sd is less, or sum_nei is less, it becomes more difficult for new water lily to emerge (calibrate based on weekly photos).*

### If cell is water lily:

*% grow*

*represented by the increase of age*

*% mortality*

*reason 1: age > lifespan*

*reason 2: external factors such as temperature and temperature difference:*

*e.g. temp<5, die out;*

   *temp<13 and temp>10 and temp_diff<-3 and sum_nei<=2*

   *=> lily on this cell dies*

*other rules based on temperature changes are calibrated based on weekly photos.*

The following flowchart gives a brief description of the model:

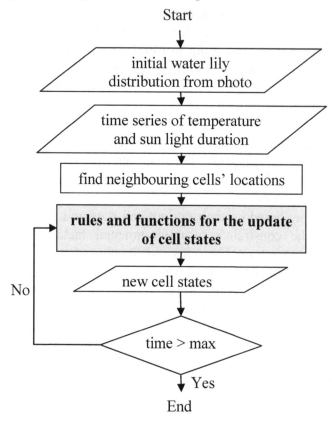

## III   Multi-Agent System modelling for two submerged macrophytes in Lake Veluwe

This model is designed for the growth and competition of two specific submerged macrophytes: *Chara aspera* (*Cs*) and *Potamogeton pectinatus* (*Pp*) existing in Lake Veluwe. The implementation of this model is shown in the flow chart below:

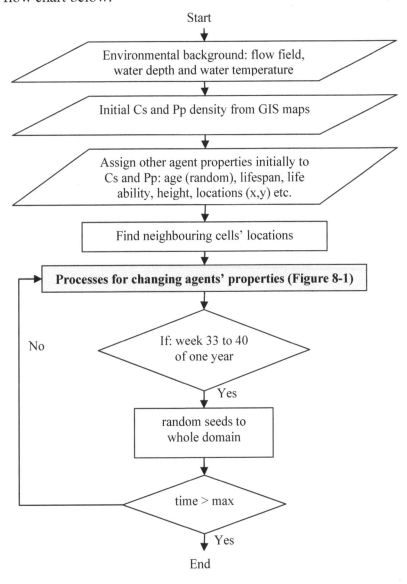

The Figure below shows the relationship between population growth and growth rate with logistic growth function shown in Eq. 7.3.

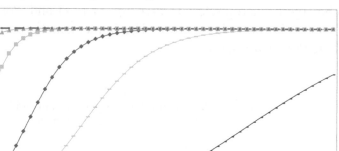

# Curriculum Vitae

Hong Li was born on 1 August 1972 in Tongliao of Nei Mongol province in Northern China, where she received her primary, secondary and high school education. She started her bachelor study in Hydrology at Hohai University, Nanjing, China in 1992 and graduated with distinction in 1996. She continued her master study in Hydrology, also at Hohai University, and obtained her MSc degree in March 1999 with the distinct thesis award from Jiangsu province. During this period, she was involved in a national special project and a project funded by Jiangsu province on coastal zone information and storm surge warning services and marine forecasts.

After graduating from Hohai University, she joined the information centre of the housing property department, real estate bureau, in Nanjing as a civil engineer. Her job was related to the development and maintenance of the Nanjing housing property information system. After having worked four and a half years for the real estate bureau, she moved to the Netherlands and entered the Master study in Hydroinformatics at UNESCO-IHE where she, graduated with distinction in 2005. She then continued her PhD research in Environmental Hydroinformatics under the supervision of Prof. Arthur Mynett as part of the joint research programme between UNESCO-IHE, Delft University of Technology and Deltares (formerly WL|Delft Hydraulics) funded by the Strategic Research & Development Department of Deltares (WL|Delft Hydraulics).

## RECENT PUBLICATIONS

**Li, H.,** Mynett, A.E., Qi, H. and Penning, E. (2009), *Multi-Agent Systems in modelling aquatic population dynamics in Lake Veluwe, the Netherlands.* Submitted to the Journal of Ecological Informatics.

**Li, H.,** Mynett, A.E. and Penning, E. (2009) *Photography-based cellular automata in aquatic plant dynamics modelling.* Submitted to the Journal of Ecological Informatics.

**Li H.,** Arias M., Blauw A. Peters S., and Mynett A.E. (2009), *Enhancing Delft3D-BLOOM/GEM for algae spatial pattern analysis: model improvement.* Proceedings of the International Conference on Science and Information Technologies for Sustainable Management of Aquatic Ecosystems, a joint meeting of the 7th International Symposium on

Ecohydraulics and the 8th International Conference on Hydroinformatics, Concepcion, Chile (Jan. 12-16, 2009).

Arias M., **Li H.**, Blauw A. Peters S., and Mynett A.E. (2009), *Enhancing Delft3D-BLOOM/GEM for algae spatial pattern analysis: Filling missing data in RS images*. Proceedings of the International Conference on Science and Information Technologies for Sustainable Management of Aquatic Ecosystems, a joint meeting of the 7th International Symposium on Ecohydraulics and the 8th International Conference on Hydroinformatics, Concepcion, Chile (Jan. 12-16, 2009).

Qi H., **Li H.**, and Mynett A.E. (2009), *Exploring Multi-agent systems in aquatic population dynamics modelling*. Proceedings of the International Conference on Science and Information Technologies for Sustainable Management of Aquatic Ecosystems, a joint meeting of the 7th International Symposium on Ecohydraulics and the 8th International Conference on Hydroinformatics, Concepcion, Chile (Jan. 12-16, 2009).

**Li H.**, Mynett A.E., Arias M., Blauw A. & Peters S. (2008), *A Pilot study for an enhanced algal spatial pattern prediction using RS images*. Proceedings of the 16th IAHR-APD Congress of Asia and Pacific Division of the International Association of Hydraulic Engineering and Research and 3rd IAHR International Symposium on Hydraulic Structures, Nanjing, China (Oct. 20-23, 2008).

**Li H.** (2007) *Scale coupling in algal dynamics modelling using DELFT3D with Fuzzy Inference Complex Automata*. Proceedings of the $32^{nd}$ IAHR Congress, International Association of Hydraulic Engineering & Research. Venice, Italy.

**Li H.**, Mynett A.E., Huang B.Q., and Hong H.S. (2007) *Main factor selection in Harmful Algal Bloom Prediction with a case study for Western Xiamen Bay*. Methodology in Hydrology; Proceedings of the $2^{nd}$ International Symposium on Methodology in Hydrology, Nanjing, China, October–November 2005. IAHS Publication number: 311, pp. 345-351

**Li H.**, Mynett A.E., and Corzo G. (2007) *Model-based training of Artificial Neural Networks and Cellular Automata for rapid prediction of potential algae blooms*. Proceedings of the 6th International Symposium on Ecohydraulics, Christchurch, New Zealand (February 2007).

Mynett A.E., **Li H.** and van Griensven, A. (2007) *Integrating flexible instruments for modelling spatio-temporal algal population dynamics:*

*linking processes and scales.* Proceedings of the 6th International Symposium on Ecohydraulics, Christchurch, New Zealand (February 2007).

Chen Q., Morales-Chaves Y., **Li H.** & Mynett A. E. (2006) *Hydroinformatics techniques in eco-environmental modelling and management.* Journal of Hydroinformatics, Vol. 08-4, pp: 297-316.

**Li H.** & Mynett A.E. (2006) *Modelling of algal population dynamics using cellular automata and fuzzy rules.* Proceedings of the 7th International Conference on Hydroinformatics, Nice, France

**Li H.**, Mynett A.E., Huang B. & Chen Q. (2006) *Harmful Algal Bloom Prediction using Data-Driven modelling: a case for Western Xiamen Bay of China.* Asia Oceania Geosciences Society 3rd annual meeting, Singapore

**Li H.** (2005) *Harmful Algal Bloom Prediction: A case study for Western Xiamen Bay, China.* MSc Thesis, UNESCO-IHE, Publ. No. WSE-HI 05-02.

## OTHER PUBLICATIONS

Wang C, Feng X, Dai J, **Li H.** (2000) *Study on surveying and mapping housing property based on GIS: A case study based on Nanjing housing property survey system.* J. Application research of computers, No. 9, pp. 85-87 (in Chinese)

Zhang X, Luo J, Chen L, & **Li H.** (2000) *Zoning of Chinese flood hazard risk.* Journal of Hydraulic Engineering, Vol.278, No.3, P1-7 (in Chinese).

Zhang Y., **Li H.** (2000) *Development of management system of coastal zone information and sub-system of storm surge warning service, Marine forecasts,* Vol. 17, No. 2, P. 64-72 (in Chinese)

**Li H.** (1999) *Preliminary research on Jiangsu coastal zone GIS and storm surge early warning system.* MSc thesis in Hydrology and Water Resources, Hohai University, China (in Chinese)

Wang W, **Li H.** & Han S (1998) *Discussion on the methodology of applied desktop GIS development.* Remote sensing information, No.3 (in Chinese).

T - #0050 - 071024 - C0 - 254/178/17 [19] - CB - 9780415558976 - Gloss Lamination